The Mackenzie River:

yesterday's fur frontier, tomorrow's energy battleground

The Mackenzie River:

yesterday's fur frontier, tomorrow's energy battleground

James K. Smith

Gage Publishing · Agincourt

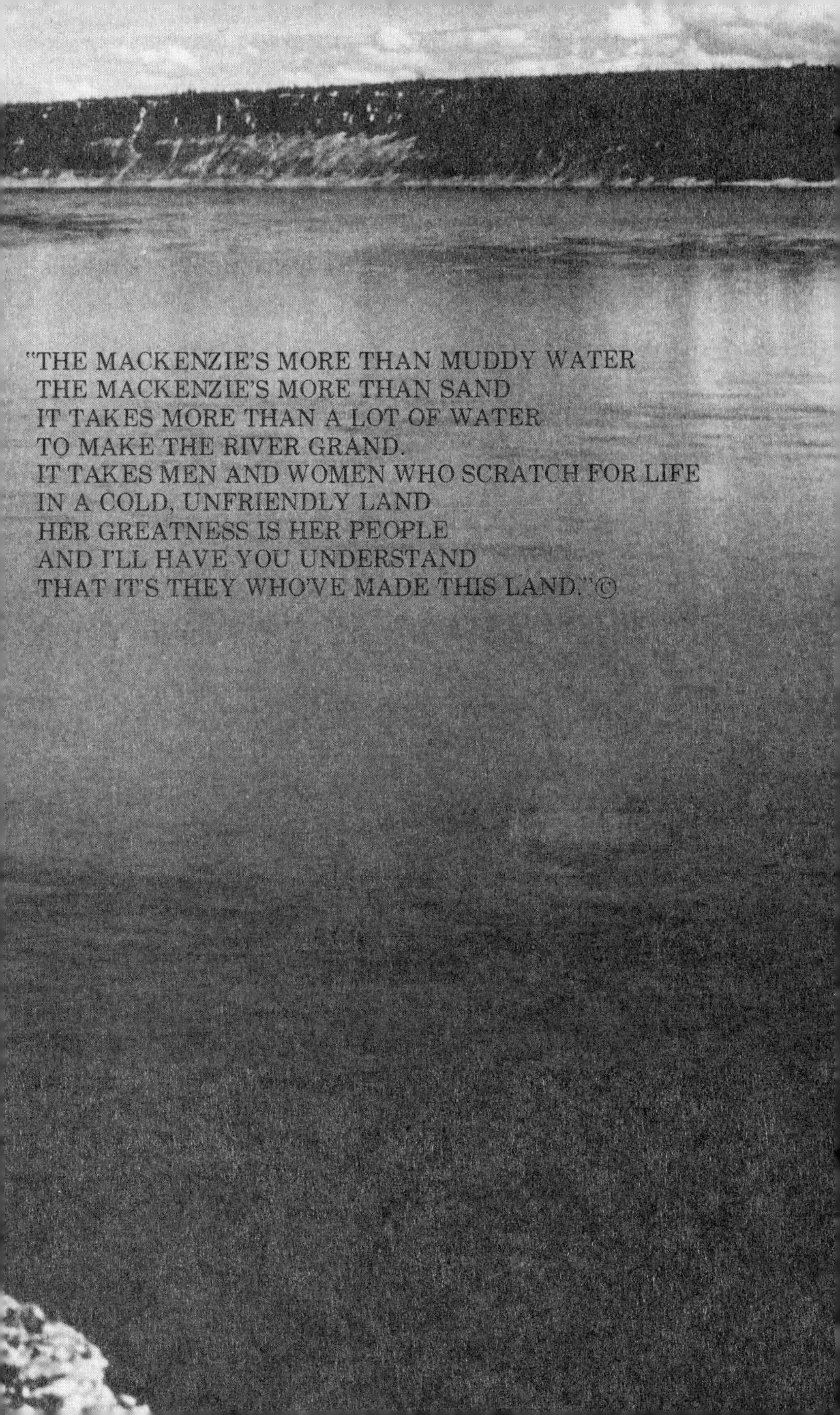

"THE MACKENZIE'S MORE THAN MUDDY WATER
THE MACKENZIE'S MORE THAN SAND
IT TAKES MORE THAN A LOT OF WATER
TO MAKE THE RIVER GRAND.
IT TAKES MEN AND WOMEN WHO SCRATCH FOR LIFE
IN A COLD, UNFRIENDLY LAND
HER GREATNESS IS HER PEOPLE
AND I'LL HAVE YOU UNDERSTAND
THAT IT'S THEY WHO'VE MADE THIS LAND."©

Canadian Cataloguing in Publication Data

Smith, James K., 1926-
The Mackenzie

Includes bibliographical references and index.

ISBN 0-7715-9980-3

1. Mackenzie River, N.W.T.
2. Mackenzie, N.W.T.-
History. I. Title.

FC4194.5.S55 971.9'3 C77-001509-3
F1100.M3S55

Design by: Fortunato Aglialoro

Jacket: N.F.B. Photothèque, George Hunter, 1963

Title page: photograph courtesy of Father René Fumoleau, OMI, Yellowknife, N.W.T.

Map by Kennedy Art Studio Limited.

1 2 3 4 5 GP 81 80 79 78 77

Printed and bound in Canada

Contents

		page
Introduction		1
Chapter 1	The "River Disappointment"	5
Chapter 2	Highway to the West	26
Chapter 3	A Battleground for Furs	55
Chapter 4	A Trading Empire	92
Chapter 5	The Oilman's River	131
Chapter 6	People and Pipelines	193
Acknowledgments		246
Further Reading		249
Index		253

By the same author

DAVID THOMPSON: Fur Trader, Explorer, Geographer

ALEXANDER MACKENZIE, Explorer:
The Hero Who Failed

To Kinmond, David, Janet, and Jellybean

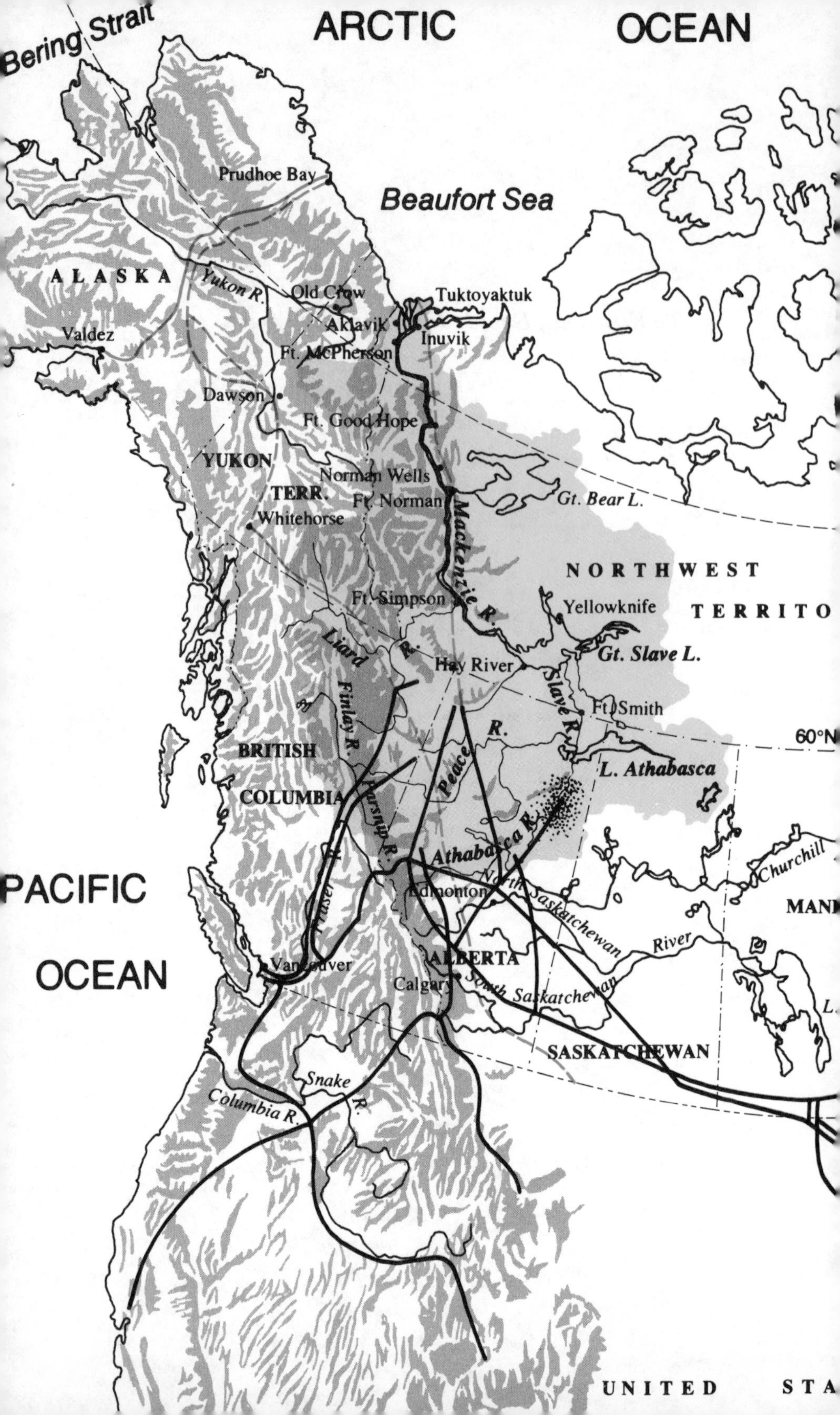

Bering Strait
ARCTIC
OCEAN
Prudhoe Bay
Beaufort Sea
ALASKA
Yukon R.
Old Crow
Tuktoyaktuk
Aklavik
Inuvik
Valdez
Ft. McPherson
Dawson
Ft. Good Hope
YUKON
TERR.
Norman Wells
Ft. Norman
Gt. Bear L.
Whitehorse
Mackenzie R.
NORTHWEST
TERRITO
Ft. Simpson
Yellowknife
Liard
R.
Hay River
Gt. Slave L.
Slave R.
Finlay R.
Ft. Smith
60°N
R.
BRITISH
COLUMBIA
Peace
L. Athabasca
Parsnip R.
Athabasca R.
Athabasca
Churchill
PACIFIC
OCEAN
Fraser
North Saskatchewan
Edmonton
MAN
River
Vancouver
ALBERTA
Calgary
South Saskatchewan
SASKATCHEWAN
Snake
R.
Columbia R.
UNITED
STA

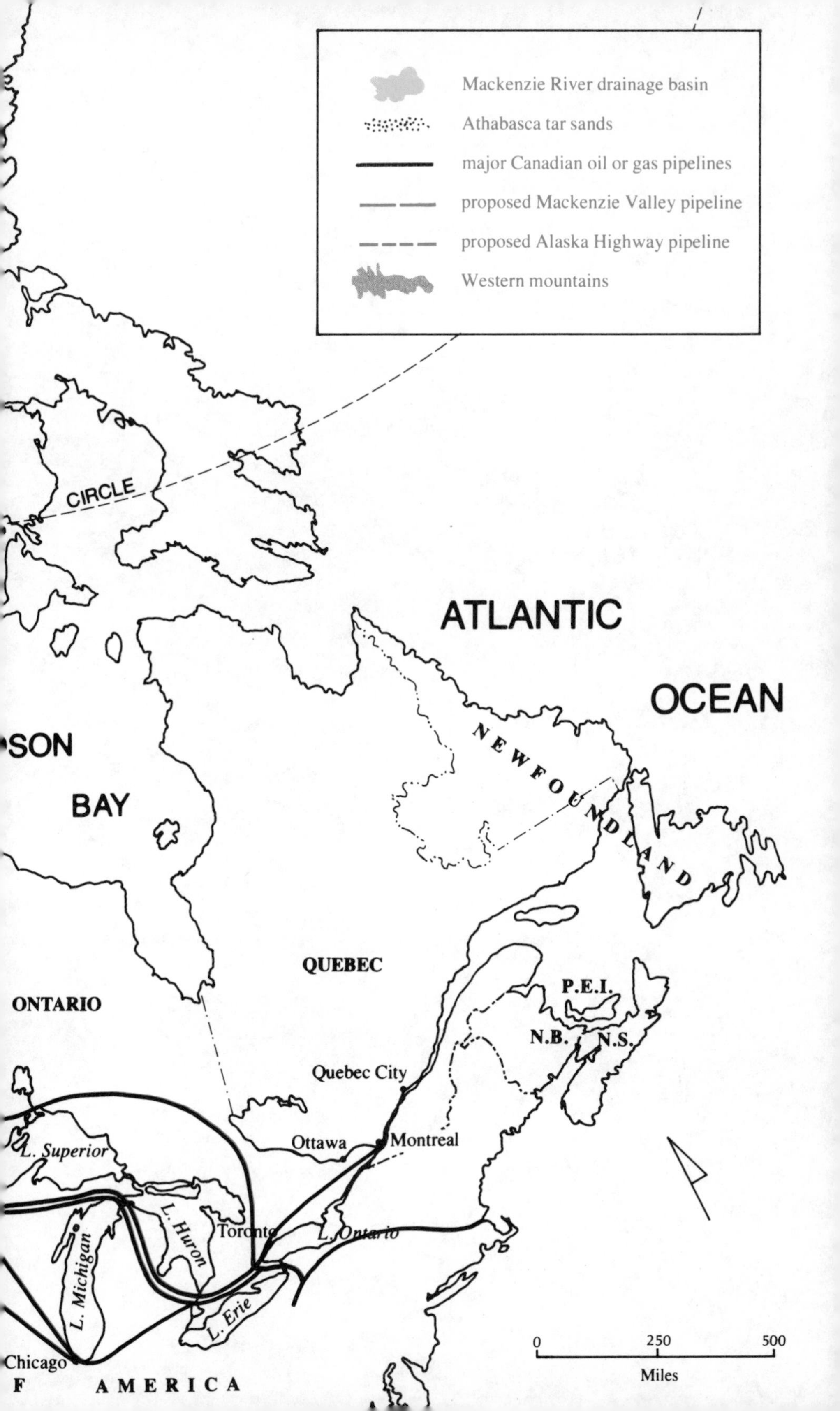
Mackenzie River drainage basin
Athabasca tar sands
major Canadian oil or gas pipelines
proposed Mackenzie Valley pipeline
proposed Alaska Highway pipeline
Western mountains
CIRCLE
ATLANTIC
OCEAN
SON
BAY
NEWFOUNDLAND
QUEBEC
ONTARIO
P.E.I.
N.B.
N.S.
Quebec City
Ottawa
Montreal
L. Superior
L. Huron
L. Michigan
Toronto
L. Ontario
L. Erie
Chicago
F
AMERICA
0
250
500
Miles

Introduction

A Canadian prime minister once remarked that if some countries have too much history, Canada has too much geography. There are few better examples of his remark than the Mackenzie River. You wouldn't think so from glancing at a map, because the river seems to start at Great Slave Lake (the one that looks like a badly-crumpled midget submarine) and end at the Arctic Ocean. Weren't we all taught that this is the waterway Alexander Mackenzie discovered and explored? Yet its 1,100 miles are only what geographers cautiously call the Mackenzie "proper," less than half its actual length. Why so? The answer is that you have to know what a river is, geographically speaking.

A river consists of a main waterway and a number of tributaries in much the same way that a tree has a trunk and assorted boughs, branches, and twigs. But the Mackenzie's trunk is very hard to detect. How many rivers have a mainstream composed of *two*, differently-named rivers – the Mackenzie and the Slave – separated by an enormous, inland sea (Great Slave Lake)? How many rivers have a mainstream that also incorporates a *second* huge body of water (Lake Athabasca)?

The Mackenzie, then, has a trunk made up of two rivers and two lakes. What about the boughs, branches, and twigs? They're many and various. But there's one great branch, the Liard, and two enormous boughs, the Athabasca and Peace rivers. Mountain-born, all three are fed by ice and snow. The Peace, however, is the most powerful tributary, and the length of the Mackenzie is calculated from the ultimate source of the Peace to the Mackenzie's mouth – 2,635 miles, to be precise.

If the Peace-Slave-Mackenzie are actually one and the same river, wouldn't it save a lot of confusion to give them all the same name? Why isn't the river called the Mackenzie all the way from its delta to the farthest source of the Peace? For the same reason the Mississippi isn't called the Mississippi all the way to its ultimate source in the headwaters of what we call the Missouri River. All these names are historic, not geographic. Somewhere back in the past, an Indian, a fur trader, or a surveying expedition gave a name to a waterway, often without any real notion of its relationship to another river or rivers. And the names stuck.

For some reason or other, the geographic principle of measuring a river from its farthest source to its mouth is usually applied in encyclopedias to the Mackenzie, but rarely to the Mississippi. Look up the former and it's normally described as 2,635 miles in length and "the second longest river in North America." Look up the Mississippi – which *is* the longest North American river – and the result is an exercise in frustration. The mileage commonly given is 2,350. So the Mississippi can hardly be the longest river in North America, can it? (What most encyclopedias don't bother to tell you is that they're describing the historic Mississippi, the river that rises in northern Minnesota and meets the Missouri to form the Mississippi "proper." At the time the headwaters of this Mississippi were discovered, the existence of the Missouri was known but its enormous extent was not even suspected.) The Missouri, then, must be the largest? Yes and no. Depending on which reference work you happen to consult, it's anywhere from 2,300 to 2,900 miles from source to its junction with the Mississippi. Several encyclopedias compound confusion, first, by citing the Missouri-Mississippi as the largest river, and then by supplying mileages of approximately 3,550, 3,710, and 3,980.

Obviously, length is not a very reliable guide to a river's size. Here again, there's another geographer's trick to be learned. What does have something to do with a river's "greatness" is the basin or area a river system drains. In other words, the whole tree. That, and the volume of water – as near as can be estimated – it carries off. On these two counts, the Mackenzie is awesome. It drains an area twice the size of Ontario or British Columbia and two and a half times that of Texas. In water volume, it ranks seventh in the world and is only surpassed in the Americas by the gigantic Amazon and the mighty Missouri-Mississippi system. In purely Canadian terms, the Mackenzie is even more of a stunner. Much of the precipitation that falls on Canada, together with the meltwater of snow and ice, enters streams and rivers and eventually finds a way to the Atlantic, Pacific, or Arctic oceans. A conservative estimate of Canada's water runoff is 2 billion acre-feet a year. Twenty per cent of this goes out via the Mackenzie, making it Canada's largest, as well as longest, river.

* * *

Canada's largest and longest river has already affected the world's destiny. Pitchblende from a mine at Great Bear Lake (the one that looks like a poorly-made swastika) was shipped up the Mackenzie in barges to Canadian and American laboratories: this was part of the experimental materials that ultimately produced the bomb that obliterated much of Hiroshima. However, the Mackenzie is going to influence the lives of many North Americans much more directly. One of its major tributaries, the Athabasca River, runs through a region of tar-sands deposits that are estimated to contain 600 billion barrels of petroleum. Sooner or later, today's costly methods of mining this "black gold" will be found supportable, and several decades' worth of oil will be piped to refineries in Canada and the United States. Trillions of cubic feet of natural gas are trapped somewhere below the surface of the river's delta or perhaps beneath the shallow waters of the Beaufort Sea, where the Mackenzie loses its identity. Sooner or later, commercial quantities of this fossil fuel will be found, and a pipeline will snake its way through part or all of the Mackenzie Valley. In one of history's great ironies, subarctic and arctic lands that could never support more than a few families in any one locality will power the machines and heat the factories, offices, and homes of millions of North Americans.

These developments are inevitable. Thanks to them, the

Mackenzie River will enter the North American consciousness in spectacular fashion. That's partly what this book is about. But only partly. The river has been a basic factor in Canadian history for almost two centuries. Fur traders used two of its tributaries to find a way westward to the Pacific – and in so doing gave Canada its proud claim "from sea to sea." In a country where waterways, free-running or frozen, were long the only highways, the Hudson's Bay Company utilized the river to operate what amounted to a state within a state – unknowingly preserving ownership of western and northern Canada until a day in 1870 when the new nation inherited about a quarter of the entire North American continent. Today, the long valley of the Mackenzie is becoming known to Canadians and Americans as a magnificent, still pristine wilderness, one of the last in North America. It has not yet been despoiled by the developers. Its Dene, Métis, and Inuit inhabitants are fighting hard to keep it that way, although the governmental and industrial-commercial odds are stacked against them.

As a matter of fact, this book is as much about people as it is about the Mackenzie River.

Chapter 1

The "River Disappointment"

Alberta's Icefields Parkway, the highway linking the Rocky Mountain resorts of Banff and Jasper, is world-famous as an avenue of towering peaks and tumbling glaciers. Roughly midway between the two towns is "the Mother of Glaciers," the Columbia Icefield. A steadily-shrinking remnant of the ice ages, it is still three thousand feet thick in places and imprisons several mountains within its 150-square-mile mass. Spilling down close to the highway is one of the Icefield's many glaciers, a six-mile-long tongue of jumbled, wrinkled, greyish-white ice. At the tip of the tongue is a small, dun-colored lake. You can drive a few hundred yards of side road to the lake's edge and park. (This is one of the few places in the world where you can step out of your car alongside a glacier.) Here, you see exactly what much of North America looked like towards the end of the most recent ice age – a natural junkyard. The slowly retreating Athabasca Glacier has dumped huge ridges of gravel all over the place, together with mounds of boulders and clumps of stones. Rock chips litter the ground, but there's not a scrap of soil to be seen. Dirt, yes, but soil, no. On the warmest, apparently windless summer day, the air has a heavy chill to it and slides off the glacier and on past you in a steady blast. Above the lake rise icy slopes. Every now and again, a glistening sheet or chunk breaks off. Everywhere there's the sound of dripping, trickling, running water.

Each summer and fall, hundreds upon hundreds of rivulets and streams pour down the frozen flanks of the Mother of Glaciers. About five miles south of the Athabasca Glacier, they form rivers that ultimately empty into Hudson Bay. The western slopes of the Columbia Icefield feed rivers that run all the way to the Pacific. But the meltwater dripping from the Athabasca Glacier is the infant Mackenzie on its way to the Arctic Ocean.

Alongside the highway leading north to Jasper, this milky meltwater skips and jumps along as the Sunwapta River. Elsewhere in the same valley, sister streams are emerging from other glaciers and swishing down sun-warmed slopes. As these waters join forces, the strength of what is now called the Athabasca River rapidly builds up in tossing, foaming billows. About forty miles north of the Columbia Icefield, the valley floor "drops" over a tall, step-like structure. Here, the Athabasca has torn a passageway out of solid rock and smashes down into its self-made canyon, sending bursts of lazy mist high into the air above the skinny pines that teeter on the canyon's edge. After this powerful display of youthful exuberance, the river takes on a calmer appearance. An ever-darkening surface is the only sign of power. As the Athabasca gouges out a deeper channel, tons of fine rock débris and sand are being picked up and kept whirling and spinning in huge, cloudy formations; in its watery depths, boulders and stones are being sent tumbling end over end as if they were mere pebbles. At its edges, the river steadily munches away at the banks, gulping chunks of earth, and sometimes gnawing off whole trees. Other mountain rivers are absorbed and vanish with little or no trace of ever having existed. Swiftly the river rolls on, creating temporary islands of silt in midstream, or casually throwing banks of debris up against its edges, where pine and fir take quick root.

The Athabasca owes its existence to glaciers twice over. Thousands of years ago, a monstrous tentacle of ice crept north to where Jasper now stands and then angled off to the northeast. It ground flat every ridge and hill in its path. It scraped, cracked, and split off anywhere up to half the mass of any mountain on its flanks. This is how the wide, flat-bottomed valley of the Athabasca was formed. Many rivers don't find their own way through or around the tall, limestone ranges that mark Alberta's western boundary. But, thanks to that ancient, glacial bulldozer, the Athabasca does.

Flanking the Rocky Mountains to the east are the foothills, a rugged, heavily-forested countryside of sandstone ridges, with lush grasslands in its broader valleys. Through this region the river follows its glacial track, but is now showing signs of age. The eager young Athabasca that started out creating its own spillway has become a mature, slower-moving waterway. Occasionally, the river roars protestingly as it is bounced down a series of rapids. But more and more of its journey is a dignified stroll through dry, plains country and then on into the great, greenish-black northern forest of spruce, birch, fir, and jackpine. The main river channel wanders in huge, casual bends from one bank to the other. Just before the Clearwater empties itself into the muddy mainstream, the Athabasca swings almost due north. Its next seventy or eighty miles are a geologist's paradise. Tar oozes from countless cracks in the crumbling limestone banks. Huge seams of coal, ugly streaks of rusty ochre, and well-defined layers of fine sand are clearly exposed. Beyond the banks, natural gas escapes from a thousand and one holes and crevices; several cauldron-like springs heave, bubble and spit out the rotten-egg smell of hydrogen sulphide.

The river flows on more and more slowly, its huge load of sediment forming hundreds of shifting sand bars and malodorous mudbanks. A little farther north, the Athabasca loses its identity in the watery confusion of a huge, marshy delta – hence the river's name, the Cree term for "the place where there are reeds." Beyond this silt-choked maze of blind channels, shallow lakes, and sloughs spread the now grey-green, now blue-grey waters of Lake Athabasca, the "top" of the Mackenzie's "trunk."

Directly across the lake from the delta is the Rivière des Rochers. Oddly enough, it used to be a two-way river. The normal flow of the Rocher River is north out of Lake Athabasca and into a larger river called the Slave. However, until the hydro-electric development of recent times, every spring season an enormous flood of dirty-brown water came surging south along the thirty miles of the Rocher and into Lake Athabasca. This annual overflow was caused by that other great source of Mackenzie water, the Peace River.

You can step out of your car and stroll a few paces to the glacial beginnings of the Athabasca. But there's no way to see the source of the Peace, Thutadé Lake, unless you charter a float plane or hire a canoe and travel into the mountains of northeastern British Columbia. Since Thutadé is 3,625 feet above sea level, an airplane is much the more painless way of getting there.

Thutadé is only 200 miles from the British Columbia coast, yet a snowflake or raindrop falling on this boomerang-shaped lake has to travel a long, long way before meeting the salt waters of the Beaufort Sea at the Mackenzie's mouth. That flake or drop not only undergoes a long descent but also suffers a rough passage. There are, of course, falls and rapids. However, in several gorges, the river smashes against one wall to ricochet off and splatter itself all over the opposite one. Before the completion of the enormous W.A.C. Bennett Dam in 1968, the most horrifying example was the Peace River Canyon a little way west of Fort St. John, B.C. In the course of 25 miles, the Peace dropped almost 250 feet, and a maelstrom of mad water raced wildly between rock walls at least 1,000 feet tall.

You'll get some idea of this tortured journey from the fact that the Peace is the only river to flow right *through* the Rocky Mountains. Other rivers in the Rockies, whether in Canada or the United States, flow east or west around this, that, or the next range. Not the Peace. It has ripped its own route out of solid rock because the river existed before the Rockies were born. When they were created millions of years ago by a very slow, upward crumpling of the earth's crust, the Peace simply tore its own trench out of the rising rock and kept on surging eastward.

Until 1968, something of this primal force could still be seen in the valley of the Peace at breakup. Each April, the four-foot layer of river ice that had lain rigid and silent since freeze-up seven months before began to show signs of life. Dull, thunderous rumblings filled the crisp, clear air, and the river's surface occasionally trembled and heaved. Far upriver, meltwater was flooding down mountainsides, finding its way via hundreds of creeks into the swelling river and prying at its icy sheath. Louder and louder became the rumblings. Every now and then, there were earsplitting cracks as the sheath shattered into crazy, jigsaw patterns. Suddenly the rumbling was so loud it could be felt as shock waves. One minute, the ice stretched hard and flat as far as the eye could see: the next, it was sundered into jumbled bits and pieces. Noise built into a terrifying roar as the Peace broke loose. It raged along, a scummy flood of water littered with debris: ice blocks the size of houses; whole stands of willow, alder, and aspen uprooted by the deluge; masses of driftwood; and the bodies of moose, deer, or bear, trapped under the ice during its formation or perhaps caught scrambling across the river's surface as it was torn from under their frantic feet. Sometimes the crest of the flood was halted in its path. The river slowly came to a standstill, rubbish piling up, layer by layer, to form a huge dam in midstream. Backing up, the Peace rose higher and higher until it overflowed the banks. But nothing on earth can resist the pressure of pent-up water indefinitely. Sooner or later, there was a rippling shudder, and the whole clogged mass staggered on downstream. Whatever held things up had just given way. If an island, it had ceased to exist – shaved off as by a giant razor. If a sharp bend, it had been rounded by the removal of several hundred tons of rock and dirt.

Far downriver near its connection with the Rivière des Rochers, the Peace is still a remarkable natural force. It rips at the clay banks and undermines them. White spruce a hundred feet tall are brought tumbling down. If one year's breakup doesn't carry them off, the next year's will. Eternally broadening its track, by the time it reaches the Slave River, the Peace is all of a mile wide.

To say the Peace "reaches" the Slave is true, although there's no real sense to the statement. The Peace and the Slave are one and the same waterway. In 1771-1772, Samuel Hearne of the Hudson's Bay Company (HBC) made a great, looping sweep of exploration overland from the Bay to the Arctic Ocean and back. On the return leg of the journey, Hearne's Chipewyan guides led him across a vast lake to a river leading into it from the south. They told him the "Slavi" tribe lived in the region; thus Hearne called the lake

"Great Slave" and the river the "Slave." These names have been used ever since. It's much the same story with the headwaters of the Athabasca River. The Sunwapta is the largest tributary. Unlike the others, it emerges directly from the Athabasca Glacier, so it has a claim to bearing the name of the main river. But the geologist-explorer who entered the Athabasca's valley in 1892 did so, not by following the river upstream, but by climbing through the mountains on the valley's eastern side. When he stumbled upon a waterway he thought far too small to be the Athabasca – and knowing nothing about the pattern of headwaters – he gave it the Stoney name of Sunwapta, "turbulent waters." In early maps of the region, the river appeared with this name. No one has bothered to change it to this day.

At any rate, the Peace, now labelled the Slave, makes an abrupt turn north. And just as the Athabasca finally spreads out into a marshy delta, so does the Slave. In a broad, flat region of marsh grasses that stretches for miles and miles, inhabited mainly by solemn pelicans and frisky muskrats, the river peters out in channels of slothful, sediment-laden water. From the air, the Slave's delta is more impressive – a chocolate-colored stain that fans out into the blue-green waters of Great Slave Lake. It was here, on a cold, blustery, June day in 1789, that a twenty-five-year-old Scot looked out across this still frozen inland sea. A fur trader by occupation, he'd just paddled down the Slave in search of a river that his employers thought would lead to salt water.

* * *

Those Canadians who have a considerable admixture of Scottish blood in their veins have good reason to be proud of Alexander Mackenzie. The trader-explorer who twice led a tiny band of men through the North American wilderness on exceedingly difficult, extremely dangerous voyages of discovery is honored as the first man to have reached the Pacific overland north of Mexico. His transcontinental crossing predates the American Lewis-and-Clark expedition by twelve years and is a Canadian "first." Even a notable, United States authority on the history of continental exploration says admiringly of Mackenzie, "In courage, in the faculty of command, in ability to meet the unforeseen with resources of craft and skill, in the will that cannot be overborne, he has had no superior in the history of American exploration."*

Like many of history's heroes, Mackenzie was a charmer. A handsome, dark-eyed Highlander, he had a stubborn mouth and chin that caused many a woman's heart to flutter. An elegant dresser and an easy conversationalist, he became a much-sought-after member of Montreal, and later London, society. He was the delight of every notable hostess and the despair of every mother with a marriageable daughter because he adored supper parties, private dances, public balls – and all the girls. (He managed to defer matrimony until he was 48.) In male gatherings, he was a cheerful drinking companion, doubtless adding much laughter to the conversation with his dry, often droll, remarks. His book, *Voyages from Montreal*,† was a bestseller, and its various English, American, French, German, and Russian editions gave vicarious excitement to tens of thousands of readers.

Mackenzie was not just gregarious. He had a canny grasp of the value of public relations. He always treated his Indian customers firmly but fairly. He habitually used a blend of fatherly sternness and kindness when dealing with his voyageurs, which is the basic reason these canoemen served him so loyally and so magnificently. As a fur-trade executive, he sought to better the salaries and working conditions of those junior to him in the trade. And the explorer who often walked casually into a village full of agitated, suspicious natives and calmed them down by giving their children gifts of sugar and trinkets had little difficulty behaving with authority and assurance in government and court circles, where he spent many years working as a fur-trade lobbyist.

There was, however, a less charming side to Mackenzie. His book, *Voyages from Montreal*, reveals a private person-

* Bernard De Voto, *The Course of Empire*, Boston, Houghton Mifflin, 1952.

† *Voyages from Montreal on the River St. Laurence* [sic], *through the Continent of North America, to the Frozen and Pacific Oceans; In the Years, 1789 and 1793. With a Preliminary Account of the Rise, Progress, and Present State of the Fur Trade of that Country*, London and Edinburgh, 1801.

ality that was distinctly egotistical. In his self-confidence, there was more than a touch of pride and arrogance. The former is most clearly seen in his personal opinions of Indians, whom he almost always dismissed outright as savages; the latter in his inability to credit colleagues and associates for guidance and help at various points in his career. Indeed, his ambition to succeed in life was so great that it generated a streak of ruthlessness that became quite marked with the passage of time and cost him at least two close friendships.

Mackenzie's basic problem was that he could lead, but he couldn't follow. He frequently ignored the need of compromise to secure the co-operation of others. For years, he sought to realize ambitions that required the support of both his trade colleagues and British politicians and civil servants. Yet Mackenzie was never quite able to win them over. His ideas were too bold: his desire to succeed, with or without help from anybody, was all too obvious. A day would come when he would be shut out of the fur trade and die an almost forgotten man.

In all justice to Mackenzie, there were factors heavily stacked against him. His principal colleagues-turned-opponents were fellow Scots as hard, grasping, and as eager to succeed in life as he was. And, fully absorbed as it was in waging a seemingly endless war against Napoleon Bonaparte, the British government just couldn't be bothered with North American matters. But it is a sad comment on Mackenzie's stiff-necked, do-it-my-way-or-else character that, within twelve months of his death, his ambitions to create a unified fur trade and to contest American domination of the entire continent began to be realized by others.

Mercifully, all this lay hidden in the future. As Mackenzie looked across the blinding white surface of Great Slave Lake that June day in 1789, he anticipated only success. He was a strong-bodied, sublimely confident young man. At twenty-five, he was already a partner in that great business combine of certain Montreal fur merchants, the North West Company. A year before, its senior directors had paid him the greatest of compliments: they'd made him manager of the Company's Athabasca Department, the richest fur region in all of North America. They had also commissioned him to find a navigable canoe route to and from the Pacific Ocean. Between Montreal and the lands surrounding Lake Athabasca lay 3,000 miles of rock, forest, lake, and river. Packs of trade goods had to be transported laboriously west, and bales of furs had to be hauled just as laboriously all the way back east again. Mackenzie had been instructed to find a western water route out for furs and in for goods. The Pacific had to be only some miles farther on beyond Athabasca – or so the Company thought. With a canoe route to and from the Pacific, the punishingly-high costs of a 6,000-mile round trip would plummet when the much cheaper method of deep-sea freighting was used between the fur markets of London and some point or other on the Pacific coast of North America. All Mackenzie had to do was find a river that was reported to flow out of Great Slave Lake and undoubtedly ran west to the Pacific. Once found, fame and fortune would be his for the taking.

* * *

Mackenzie's accounts of his two voyages are written in lean, laconic prose. It doesn't exactly entertain a modern reader, but it certainly informs. His keen eyes missed little. An exceedingly meticulous Scot, Mackenzie more or less took inventory as he travelled. In the *Voyages from Montreal*, his journal entry the day he arrived at Great Slave Lake is a typical example.

> Tuesday, June 9. We embarked at 1/2 past 2 a.m. Calm and foggy. Soon after, 2 young [Indian] Men joined us that we had not seen for 2 days. They [had] killed 4 Beavers and 10 Geese which they gave me. Our course N.W. by N. 1 mile . . . and steered S.W. by S. 1 1/2 miles, W.S.W. 1 1/2 miles, W. 1 Mile. Here we entered a small river on the East side [the Jean River, a branch of the Slave River emptying into Great Slave Lake just east of the river's main channel]. They tell me there used to be a Carrying Place [portage] at the Entry owing to Drift Wood filling up the Passages, but the Water has carried it off. The course of it is winding but [I] suppose it will be about North distance 10 miles to where it falls into the Slave Lake, and where we arrived at 9 a.m. We found a great change in the Weather, it being excessive Cold. The Lake is covered all over

with Ice and does not seem to have yet moved, excepting along the shore. All along the River we were much troubled with Muskettoes and Gnatts, but now . . . they must take their leave of us till the Weather gets warmer. From below the Rapids [farther back up the Slave], as well as above, the Banks are well covered with all kinds of Wood peculiar to this Country, particularly the West Side, the last being lower and a richer soil. On the East Side, the banks are high, the soil is yellow Clay and Sand, so that the Wood is not so big nor so numerous. The ground is not yet thawed above 18 inches, notwithstanding the Leaf is at its full Growth. The Indians tell me that at a very little distance on both sides [of] the River are very extensive Plains, where there are vast herds of [Wood] Buffaloes, and that the Moose Deer and the largest kind of Rein Deer [caribou] keep in the Wood close by the River. The Beaver, which are numerous, build their Houses in small Lakes and Rivers, which they cannot do in the larger River [the Slave], as the Ice carries everything along with it in the Spring. All the Banks of the River are covered with wild Fowl. We killed 2 Swans, 10 Geese, 1 Beaver this morning without losing an hour's time, so that if we were [out] for the purpose of hunting, we might soon fill our Canoe.

He was a bit optimistic about the mosquitoes. Three days later he was forced to write in his journal, "Towards noon our old Companions, the Muskettoes, visit us in greater Numbers than we would wish as they are very troublesome Guests." They were nothing if not companionable. They would pester, pursue, irritate, and infuriate the explorers all the way down to the Arctic and back again. North West Company canoemen were accustomed to early starts, but Mackenzie and his men would sometimes cut short a night's rest just to get out on the water away from these terrible tormentors. Later on in this voyage, Mackenzie halted to climb to the top of what he called a "prominent hill" and view the surrounding land. Accompanied by two voyageurs and his Chipewyan interpreter-hunters, it took an hour and a quarter to ascend what proved to be a fair-sized mountain. They had barely time to get their breath back and look around when "we were obliged to shorten our Stay here on account of the Swarms of Muskettoes that attacked us and were the only Inhabitants of the Place." On another occasion, when he and a Chipewyan again attempted to climb a high hill, "We were obliged to relinquish our design halfway up it, being nearly suffocated by swarms of Muskettoes." It is characteristic of Mackenzie that he makes more of encounters with mosquitoes than he does of such espisodes as his canoe capsizing in white waters or, on one occasion, nearly being stabbed to death by an angry Indian.

Great Slave Lake caused the explorer a great deal of trouble. He had arrived at its southwestern end, an area where the prevailing winds drive a winter's accumulation of ice against the sandy southern shore and often right up on to it in massive, glistening piles. So he couldn't advance any farther northward. And shifting ice often made it impossible to net fish in the shallows. Yet Mackenzie had to ensure that there was a plentiful supply of food for five voyageurs – two of whom had their Indian wives along – the chief guide and his two wives, two other Chipewyans, and Laurent Leroux, one of his traders. Even after several days of drenching rainstorms and blustery winds broke up the ice, his problems were far from over. About 30-50 miles wide and almost 300 in length, Great Slave is as tricky to cross as any stretch of open sea and is often swept by high winds. The lake is dotted with islands of every size and shape. Some of them barely rise above the water, while others tower as high as 600 feet with sheer cliffs for edges. It took Mackenzie's party a week of dodging from one island stepping-stone to the next to reach the rocky north shore. Somehow, his men managed to prevent their frail canoes from being crushed by rampaging ice floes or ripped open by sheet ice, which formed rapidly even in the month of June.

On any map, Great Slave Lake looks like a badly crumpled submarine that is just about to make its last dive. The conning tower is formed by the modern Yellowknife Bay. Mackenzie's party hurriedly landed on its southeastern edge. Here, they met three lodges of, in his words, "Redknife Indians, so called from their copper knives."* One of them was sent off to fetch the members of other lodges living nearby. As it turned out, while these natives had beaver and marten skins to trade, they could not give Mackenzie what

* They have also been referred to as Copper Indians and Red Knives. But Yellowknives is the common modern term applied to the descendants of this particular Indian group.

he really wanted – detailed information about a river that he'd been told emptied out of the lake's western end. However, one of the Red Knives thought he had seen the beginnings of such a waterway and was immediately engaged as the party's guide. Ever the profit-minded Scot, Mackenzie lectured his customers sternly on the benefits they would derive from trapping and promised them a trading post, which would be maintained "as long as they would deserve it."

At 2 o'clock on the morning of June 25th, the senior canoeman roused his fellows with the traditional reveille of "*Lévé! Lévé! Il faut partir!*" Blanket-wrapped voyageurs rolled out from under their upturned canoes. Within minutes, these were riding low in the water, and the men waited impatiently in the shallows for the command "*Embarque!*" In the normally calm, early hours of the morning, they could make a lot of progress and would then be permitted to stop for a bite of breakfast. But this particular day Mackenzie was taking a little longer than usual to get started. He was leaving Laurent Leroux behind to spend the summer working the north shore for furs and wanted to impress on him the importance of drumming up as many customers as possible. Mackenzie also had to check his field reports. These would be taken back to his cousin, Roderick McKenzie, whom he had left in command of the Athabasca Department's day-to-day operations during his absence that summer. In turn, these would be forwarded from Fort Chipewyan on Lake Athabasca to headquarters in Montreal. While he was about to venture into territory unknown to any fur trader, business was still business. His comments on expanding the activities of the Department northward would be just as important to the Company as finding some river that was supposed to link up with Pacific tidewater.

By 3 a.m., he was ready to "march," as fur traders always called canoeing. The voyageurs and Indian employees staying behind at Great Slave raised their guns skyward "to salute us with several vollies, to which we returned a few shot." As the explorers paddled westward across the mouth of Yellowknife Bay, ragged curtains of mist quickly hid their friends from view.

It took Mackenzie four days to find his river, days spent warily circling past drift ice or hastily running ashore until treacherous banks of fog cleared away. To add to his impatience, the "Red Knife" turned out to be a poor guide. On one occasion, he directed the explorers into a deep bay, which he thought might be the entrance to the river. But its waters lacked any clear sign of a current and finally brought them up against masses of broken ice, whose presence was made even more menacing by patches of mist. Back paddling very carefully, the party found refuge for the night on a nearby island. A similar disappointment had to be endured the following day. This time the guide assured them they would arrive at what he called the "Grand River." Yet, hours later, the canoes scraped to a halt in shallow waters. Whereupon, "English Chief," the leader of Mackenzie's Chipewyans, became utterly enraged and wanted to shoot their guide. Mackenzie doesn't say whether or not fear improved an imperfect memory, but at that very moment the Red Knife suddenly recalled journeying from the river through wooded country to the very spot where the canoes were now beached.

He was right - or a lucky guesser. Breaking camp at four the next morning, the party backtracked out of the bay, rounded yet another headland, and soon detected a strong, swift current taking them out of Great Slave Lake. Some hours later, a stiff breeze came out of the east. Hoisting sails, the explorers moved along at a very brisk rate.

At this point in his journal, the habitually matter-of-fact Mackenzie says nothing of his feelings. Yet he must have been elated. Having fixed Great Slave Lake as being at 61° 40′N - approximately the same latitude at which, eleven years earlier, Captain James Cook of the Royal Navy had noted a large river emerging from the interior of North America into the Pacific Ocean - he was now on a major waterway leading westward. It just had to be the one Cook described in his book as thick, muddy water "very considerably fresher than any we have hitherto tasted" and full of "large trees and all manner of dirt and rubbish." It had to be the same waterway that Captain Cook had called a "great river" and which had been officially named "Cook's River" in his honor. Unfortunately, like many a man before and after him, Mackenzie was bending facts to fit a theory. There is no water connection between Great Slave Lake and the Pacific. "Cook's River" was nothing more than a complex of minor waterways that create an estuary now known as Cook Inlet. (At the eastern end of the inlet stands the modern city of Anchorage, Alaska.)

Four days later, Mackenzie calculated they had run downstream 217 miles west and 44 north. The satin-slick current, which "makes such a hissing . . . as a kettle moderately boiling," quickened to eight or ten knots in some stretches. The next day the river widened, the current slackened a bit, and he caught sight of a ridge of high snowy mountains ahead. He was sure that, sooner or later, the river would pierce these mountains and lead to the Pacific. There is only one faint suggestion of disappointment in his journal at this time: when the weather warmed up from time to time, he was pestered by buzzing, biting squadrons of his "old Companions."

But as day succeeded day, the mountains always remained tantalizingly at a distance to the west – as they would do all the way to the Arctic Ocean. Like a long row of gigantic, white-capped sentries, they silently warned him to go wherever the river took him. The only point at which he might have slipped between them was the entry of that other great Mackenzie tributary, the Liard. Some 750 miles long, it rises in the Yukon, races south into British Columbia, and then doubles back into the Northwest Territories. But the explorer never saw the Liard. He knew about it from native report, but missed it. On the morning of July 2nd, he had delayed starting off until 5.30 a.m. due to thick fog. By 7 a.m., when it cleared, he suddenly noticed that "the water has changed from being limpid to muddy, which we suppose must be some large river falling in from the southerd [*sic*]." Luckily, he didn't realize the significance of this muddy water. The Liard would only have led him into a maze of headwaters walled in by several mountain ranges.

A week after entering the river, the strength of the current had whisked the party almost halfway to the Arctic Ocean. (Mackenzie continually underestimated the water's force; he averaged close to 100 miles a day going downstream.) In the neighborhood of modern Fort Norman, Mackenzie had his first encounter with local residents and received a faint hint of where he was heading:

> Sunday, July 5. . . . There are five Families of them in all, 26 or 30 Persons, and of two different Tribes, Slave and Dog Rib Indians. . . . The information they gave us respecting the River seems to me so very *fabulous* [Mackenzie's italics] that I will not be particular in inserting [it]. Suffice it to say that they would wish to make us believe that we would be several Winters getting to the Sea, and that we should be old Men by the time we return. That we would have to encounter many Monsters (which can only exist in their own Imaginations). Besides that there are 2 impracticable Falls or Rapids in the River, the first 30 Days March from us, &c. &c.

"Fabulous"? Monsters that could "only exist in their own Imaginations"? Perhaps so, perhaps not. These Indians did advise him he would reach *salt water*. And there is something rather monstrous about the polar bear and the white whale. (Mackenzie was to hear more about these animals from natives farther downstream; when he finally saw several white whales, he noted in his journal that a "stroke from the tail of one of these enormous fish would have dashed the canoe to pieces.") In any case, why would these Slaves and Dogribs encourage an obviously single-minded, generous stranger to go on downstream and hand out many marvellous gifts to all and sundry? This rich, white-faced man wanted nothing except information about the river. So why not relate the many stories they knew about it, which was the polite thing to do, anyway?

In the *Voyages from Montreal*, Mackenzie describes these informants in considerable detail.

> . . . They are all an ugly, meagre, ill made People, particularly about the Legs which are very clumsy & full of Scabs by their frequent roasting them to the Fire. Many of them appeared very sickly owing as I imagine to their Dirty way of living. . . Some of them wear their hair long, others have it long behind & from the Crown down to the two Ears cut quite short, but none of them takes any pains to keep it in order. A few of the Old Men had

> their beards long & the rest had it pulled out by the Roots [so] that not a hair was to be seen. The Men have two double lines Black or Blue tatoed [*sic*] upon each cheek from Ear to Nose. . . . They make their Clothing of the Skin of the Rein or Moos [*sic*] Deer well dressed [i.e. cut and finished] but the Skin of the first is more Common and they dress many of them in the Hair [i.e. hair side out] for Winter wearing; of either they make shirts which come down half their Thighs, some of which [shirts] they embroider very neatly with Porcupines Quills & the Hair of the Moos [*sic*] Deer painted Red, Black, Yellow and White . . . the Women dress the same as the Men. The men have no covering on their Private Parts except a small Tassel of Pairings [*sic*] of Leather, which hang loose by a small cord before them in order as I think to keep off the Flies which wou'd [*sic*] otherwise be very troublesome. Both Young & Old Men have the glans of the Penis uncovered; indeed their want of Modesty and their having no Sense of their Nakedness but from the Cold would make a Person think that they were descended from Adam . . .

He goes on in this cool, clinical fashion for several pages, for all the world as if he were visiting a rather interesting zoo. But then, elsewhere in the *Voyages*, his descriptions of Indians are just as detached and equally patronizing. Like almost every other fur trader, Mackenzie felt immensely superior to his customers and, despite the fact that he spent nine years of his life living among them, took neither the time nor the trouble to view Indians in light of their environment. He failed to appreciate that these northern people had found ways to survive, even thrive, in an often hostile land. They did not actually *need* anything a fur trader had to offer. They made their own hunting weapons, household implements, clothes, and homes. In accordance with age-old oral tradition, they were their own teachers, priests, and doctors. Mackenzie was more concerned with the appearance than the character of his potential customers. He saw only their tattooed bodies and clothes embroidered with beads, quills of goose and porcupine, and strips of moosehide. He did not realize that, together with red-ochred tents and numerous designs on common birchbark containers, these were ways to offset the grimness of a far northern environment. It's very, very easy to misconstrue and misrepresent another culture. To those who have never seen them before, a kilt and a sporran are pretty weird, even dubious, items of clothing. Is the kilted person a man, or a poor unfortunate woman cursed with a deep voice and hairy knees?

The fact that Mackenzie had little or no difficulty communicating with various Indian groups in the Northwest is not as remarkable as this might appear. All of the natives he met in the course of his voyage to the Arctic belonged – as did his Chipewyan hunters – to a common linguistic stock, the huge Athapaskan-speaking family. They have often been described as the "babiche people." Of all Indians, the Athapaskans were those most given to the use of *babiche*, strips of tanned caribou or deer skin used for lacing snowshoes, making nets, bags, snares, and bindings of various kinds. They were a nomadic people, occupying the farthest reaches of the vast, northern forest that stretches from Labrador all the way west and north into Arctic regions. They earned a living by hunting, fishing, and trapping. Essentially hunters and inured to the long, severe, northern winter, theirs was a culture that reflected an eternal struggle to find enough food and warmth to survive a harsh environment – hence spears, bow and arrows, stone or flint arrowheads and knives, snowshoes, toboggans, birchbark canoes, skin or bark containers and cooking "pots," and hide clothing. It took about ten caribou or deer hides to provide a man or a woman with a shirt, leggings, robe, moccasins, cap, and mittens. (If you think it's a smart idea to attach a string between a pair of children's mitts, give the credit to some Athapaskan man or woman.) And it took almost as many hides to equip an Athapaskan husband or wife with what they considered a plentiful supply of babiche.

Mackenzie's means of communicating with the natives he met were his Chipewyan hunter-guides. For years, the Chipewyan were the most numerous Athapaskan group in northern Canada and occupied territories extending from the western shores of Hudson Bay to the eastern fringes of Lake Athabasca and Great Bear Lake. But the widespread smallpox epidemic of 1780-1781 almost destroyed them: according to Samuel Hearne of the HBC, it killed nine-tenths of the entire tribe. An edge-of-the-woods people, the

Chipewyan lived in small bands and hunted caribou on the open, wind-swept tundra in summer and in the northern forest, where these animals wintered. Hearne, who travelled in the North with some of them for two years said they could "procure a comfortable livelihood" with just a hatchet, a knife, a file, and an ice chisel made of moose horn. For many years, they acted as middlemen between the Hudson's Bay Company and the tribes of the Mackenzie Valley until their great enemy, the Cree, acquired firearms from French traders and, later, the HBC. English Chief, the senior guide, was, according to Mackenzie

> . . . one of the followers of the chief [Mattonabee] who conducted Mr. Hearne to the Coppermine River [flowing into the Arctic] and had since been a principal leader of his countrymen, who were in the habit of carrying furs to Churchill Factory, Hudson's Bay, and till of late much attached to the interest of that company.

His countrymen's name is said to have been derived from a Cree term meaning "pointed skins," a sarcastic reference to the form in which the Chipewyan dried beaver skins.

The Copperknives Mackenzie met on the north shore of Great Slave Lake hunted caribou and muskoxen in the land between Great Slave and Great Bear lakes in summer and wintered along the Coppermine River. Here, they mined copper to make spear and knife blades and arrow heads. The "Slave" Indians moved up and down the Slave River and around the western fringes of Lake Athabasca in search of moose. In general a mild, peace-loving people, the Slave – a Cree term of contempt for people they often raided and took prisoner – were actually feared by neighboring tribes, who considered them very skilled at witchcraft. The Dogrib, so-called because of their legendary descent from a dog, roamed east of the Mackenzie, approximately midway between Great Slave and Great Bear. They had a reputation for liveliness, a great fondness for games and drum dancing. Mackenzie would also meet some of the People of the Great Hare, as timid as the Arctic animal that provided much of their sustenance and clothing, and the Loucheux (French for "slant-eyed"), or, as they called themselves, Kutchin, "The

People." Mackenzie called them the "Quarrelers," perhaps because he heard stories of their fierce fighting spirit and love of excitement. On his way to the Pacific four years later, Mackenzie encountered other Athapaskan-speaking tribes: the Beaver of the Peace River, the original name for which was Tsades, "River of Beavers"; the Sekani, the "People of the Rocks," that is, of the Rocky Mountains, who had been driven west by both Beaver and Cree into lands edging the upper waters of the Peace River; and, in the lake-and-river country of what is now central British Columbia, the Carriers, whose custom it was to make a widow carry, in a basket on her back for three years, the charred bones of her former spouse.

After spending the better part of nine hours chatting with these Slave and Dogrib families, Mackenzie took to the water again. Barely an hour later, "we passed the River of the Great Bear Lake, which appears to be a fine deep River about 100 yds wide; the Water is quite clear, of the color of Salt Water." This clear, greenish water remained distinct from the muddy mainstream for many miles. (Without realizing it, Mackenzie was looking at the outflow from one of the largest bodies of fresh water in North America, all of 12,000 square miles in area and with depths of 2,000 ft. or more.) Pushing on down the island-studded waterway between steep, sloping banks of clay and gravel, Mackenzie soon came to where the river had gouged out a trench so deep that sheer limestone cliffs towered 100-200 feet above the water for a distance of seven miles – the famous Ramparts.

As always, the canoemen were more interested in their bellies and kept eyeing the large, plump whitefish, northern pike, and Arctic grayling lying on the bottom of the canoe, presents from local Indians. But even the happy-go-lucky voyageurs began to pay more attention to the countryside on either side as successive Indian groups spoke of a hostile people they called "Eskmeaux" who lived farther downriver. One of their Indian informants displayed his knowledge of these people in a rather startling way. After performing an Eskimo dance in Mackenzie's canoe right in the middle of the river – at which the crew told him to shut up and quit rocking the boat – the fellow "before he sat down pull'd his

Penis out of his breeches, laying it on his hand & telling us the Eskmeaux name for it. In short he took much Pains to show us that he knew the Eskmeaux and their customs." As for Mackenzie's Chipewyan guides, they were scared stiff by the bleak, vaguely menacing, appearance of a landscape that was beginning to flatten out noticeably into slimy mud flats and banks.

Their imperturbable leader was, as always, busy investigating everything he could. At a deserted Eskimo summer camp, Mackenzie didn't hesitate to get down on hands and knees and crawl along a narrow passageway into an earth-and-wood home, work out its dimensions, and enter these in his notebook. In the nearby countryside, he observed the Arctic's beauty and barrenness side by side. Hills were covered with grasses and flowers, but the valleys in-between were choked with ice and even, in July, snow. "The earth is not thawed above 4 inches from the surface; below is a solid body of ice." His instruments confirmed he was just a degree or two above the Arctic Circle, which quite astonished him. Obviously the river was more powerful than it looked. In point of fact, the current acted like a magic carpet. It had wafted Mackenzie and his men 1,100 miles in 14 days.

Exactly two weeks after finding the river, everyone settled down to get what sleep they could in the glaring light of the midnight sun. Within minutes, some of the voyageurs were up again rescuing baggage from rising water – Arctic tidewater.

Actually, Mackenzie had some difficulty convincing himself that he had reached the sea. In the first place, the last 150 miles of river travel had been a matter of threading a way through a fantastic collection of lakes surrounded by innumerable, muddy water channels. (The Mackenzie Delta contains something like 15-20,000 individual lakes.) He thought he had perhaps arrived at what local Indians called "the Lake." The water tasted fresh, and there was a surprising abundance of trees, mainly white spruce; the nets set each night still produced varieties of fresh-water fish each morning. Then he began to come across unmistakable signs of salt water. One morning, the nets held "a fish about the size of a Herring, which none of us know, except the English

Chief, who says they are very plentiful at Hudson's Bay." Next day, he encouraged his men to chase what he recognized as white whales (belugas). Twenty-four hours later, the water rose about sixteen or eighteen inches during the early hours and soaked the baggage. This time, the wind had not changed or blown harder, so he decided he had to be on the edge of the Arctic Ocean, or on one of its coastal lagoons.

There was nothing to do now but backtrack as quickly as possible. The short Arctic summer was almost over. Winter might catch and imprison him and his men somewhere along the many miles that lay between the ocean and home base on Lake Athabasca, far, far away at the "top" of the Mackenzie's "trunk." The expedition had been a failure: it had not led to the Pacific. All he had done was discover what he is said to have called the "River Disappointment." But then Alexander Mackenzie had no way of knowing that he had explored and charted the North American waterway that is exceeded in length only by the mighty Mississippi-Missouri system.

Chapter 2

Highway to the West

Long before the Mackenzie River became a trade route in and out of the north, its headwaters were used as a sort of Trans-Canada Highway. Of its two great tributaries, one – the Athabasca – is a natural track that can be followed to and from the Rocky Mountains. The other, the Peace, leads right through the Rockies.

A remarkable number of people journeyed up and down these two waterways. Most of them, of course, were employed in the fur trade. But one or two were scientists, and a famous Canadian artist – Paul Kane – also made the long overland trip to and from the Pacific. Some travelers were actually tourists and belonged to the breed of Englishman that is happiest when wandering about in remote parts of the world. However, the first man west was that remarkably persistent fellow, Alexander Mackenzie.

* * *

The motive behind Mackenzie's two explorations was a strongly economic one: North West Company executives desperately wanted to find a navigable waterway to and from the Pacific coast. Given such a route, trade goods and furs could be freighted *in bulk* between Europe and the Northwest instead of being hauled at enormous expense over the many miles of wilderness waters between Montreal and Grand Portage on the north-west shore of Lake Superior and then between that depot and all the interior posts. These executives knew perfectly well that a cheap supply route by sea was the basic strength of the Hudson's Bay Company, together with that fact that its main depots on the Bay – Churchill Factory and York Factory – were farther north and west than Grand Portage. Mackenzie thoroughly agreed with this shrewd appraisal. His early years in the accounts department of a Montreal fur merchant had shown him that fur trading was financed by borrowing enormous amounts of capital in the form of trade goods. His early years trading in the wilderness convinced him how dangerous – if not insane – it was to expose this capital to the vagaries of rock and rapid. As he stressed in the *Voyages from Montreal*, transport to and from the most distant parts of Athabasca

> . . . occupies an extent of from three to four thousand miles through upwards of sixty large lakes and numerous rivers and the means of transport are slight bark canoes. It must also be observed that these waters are intercepted by more than two hundred rapids, along which the articles of merchandise are chiefly carried on men's backs and over an hundred and thirty carrying-places, from twenty-five paces to thirteen miles in length, where the canoes and cargoes proceed by the same toilsome and perilous operations.

This fantastically long, fantastically expensive transport route – to which must be added another 3,000-mile haul of goods from and furs to Britain – was the reason why the margin of profit over cost was never very great in the Montreal fur trade. In fact, Mackenzie estimated that the cost of transportation was one half the total cost of carrying on the trade. It is true that a few men made fortunes in the trade but only, as one writer has remarked, "by penny pinching and a driving of the voyageurs to a degree which would horrify a modern trade union."

It was in the course of his return from Arctic tidewater that Mackenzie first heard faint echoes of Europeans on the Pacific. The first occasion was when a band of Indians recounted an Eskimo report of meeting white men "in large canoes" far to the westward "eight or ten winters since" and exchanging leather for iron. This contact might have been with Captain James Cook, but was more likely with Russian traders somewhere on the northwest coast of Alaska. Some days later, Mackenzie talked with a Dogrib, who repeated a Hare Indian story of a mighty river on the other side of the western mountains that fell into the "White Man's Lake" far to the northwest. (At this time, Mackenzie was so far north that these references were probably to the Yukon River and the North Pacific Ocean.) The Dogrib said that the natives who lived at the river mouth made "canoes larger than ours," which tallied with Cook's descriptions of Pacific coast dugouts and war canoes; they killed "a kind of large beaver, the skin of which is almost red," which was unmistakably a description of the sea otter, whose pelt was the glossiest, if not also the hardest-wearing, of all furs.

So Mackenzie set off once more from Fort Chipewyan in search of a water route to and from the Pacific. This time, he went ranging up the Peace River, which seemed to have a long, westward-reaching course. If he was going to get to the Pacific and back in the course of a summer season, the nearer the ocean he started, the better. After wintering as far up the Peace as possible, he would be able to make a quick dash to tidewater and then hurry back to his depot on Lake Athabasca before winter closed in again.

Early that October morning, Mackenzie's tiny "brigade" of three canoes paddled across the choppy, chilly waters of Lake Athabasca. Two days later, he and his followers turned westward into the mile-wide, muddy mouth of the Peace River. Mackenzie noted in his journal that the waterway derived its name from Peace Point, a natural feature about 20 miles from the mouth of the river and "the spot where the Knisteneaux [Cree] and Beaver Indians settled their dispute." Here, the Cree, who had invaded the Mackenzie basin and driven what they called the "Slave" Indians northward and the Beavers westward, agreed upon a common boundary with the latter.

As on the trip to the Arctic, the canoes had been equipped by Mackenzie with masts and sails, which, when the winds were northeasterly, helped the party to make good headway. They had to hurry along. The weather was bitterly cold: "There were frequent changes of the weather in the course of the day, and it froze rather hard in the night. The thickness of the ice in the morning was a sufficient notice for me to proceed."

For days on end he drove his men hard. On November 1, they managed to reach the place that had been picked as a wintering quarters, a site a few miles west along the Peace from its junction with the Smoky (close to the site of the modern town of Peace River, Alberta). Waiting there were two men who had been sent ahead in the spring to square timbers and build palisades. A large number of Indians had gathered to welcome him with volleys of musket shot. Mackenzie called them to him and gave each about four inches of tobacco and a dram of spirits. Mackenzie made it quite clear that he would be hard on them if they failed to bring in the returns in furs he expected from them. To this end, he spent the next six days equipping these Indians for their winter hunting and also arranging for the provisioning of his post with fresh meat. Only when this was done did Mackenzie give his full attention to the construction of Fort Fork.

Unfortunately for the voyageurs, the first bitter breaths of winter made the labor of construction difficult and tiring. On November 27, for example, with the temperature hovering around the zero mark for the fourth successive day, little or no work could be done: the effect of the frost was such that the axes became as brittle as glass. It was almost Christmas before Mackenzie was able to move out of his tent and into the house erected for him. The men hurried to construct bunkhouses for themselves and a storehouse for the winter's take in pelts.

On the first day of January, Mackenzie was abruptly awakened at daybreak by musket shots. In a company largely run by Highland Scots, Christmas was more or less another working day, but the traders had long encouraged their canoemen to celebrate New Year's Day with gusto. Mackenzie took the hint and hastened to give them plenty of spirits and a special issue of flour with which to make cakes. The new year was only a few days old when two "Rocky Mountain Indians [Beaver Indians]" turned up at Fort Fork with the information that just beyond the mountains was a great waterway running towards the midday sun. They also informed Mackenzie that all the way to the mountains the countryside of the Peace was abundant with animals. These seemed omens of success, and the latter information was a great relief. Like every trader, Mackenzie's perpetual problem was finding enough game and fish to be able to conserve pemmican supplies against those unpredictable times when game became scarce or disappeared completely.*

On the evening of May 9, 1793, Mackenzie set off west up the Peace River. He took with him as his second-in-command, Alexander McKay, a Company clerk, six voyageurs – two of whom, Joseph Landry and Charles Ducette, had accompanied him to the Arctic – and two Indian guides. They all travelled in a canoe built to Mackenzie's specifications. It was large enough to carry the ten men (and a dog) and three thousand pounds of provisions, goods, ammunition, and baggage, yet on a good portage two men could carry it for three miles without resting.

For the next few days their surroundings were very beautiful. Rising gently from a river front edged with alder and willow, the grassy valley of the Peace was almost lawn-like in places and dotted with groves of poplar. As the two Beaver Indians had said, it was rich in animal life, particularly buffalo and elk, which wandered through the region in vast herds. About a week later, at which time the explorers were just a little west of the present-day town of Fort St. John, British Columbia, all of them were cheered to see the Rocky Mountains appear to the westward, their summits shrouded in snow. Things were going well. Paddling or poling, the men were making between 10 and 20 miles a day against the surging, spring-swollen waters of the Peace.

Forging on upriver, they came to a location that natives had warned Mackenzie was a succession of rapids, cascades, and falls. These were always bypassed in the course of a laborious portage over the mountainous north flank of the miles-long Peace River Canyon. Had he taken this route from the start, Mackenzie would have avoided a great deal of toil and trouble. But he didn't realize what he was letting himself in for and directed his men into the canyon itself.

After towing the heavily loaded canoe for about a mile, the party was forced by overhanging rock to cross over to the north side of the Peace, where stones loosened by spring

* Pemmican was made from pounded buffalo, deer, bear, or elk meat mixed with animal fat, all of which was sometimes flavored by the addition of berries. One of the first "instant foods," it was filling and sustaining rather than appetizing.

thaws and recent rains were continually slipping or rolling down into the river. The bank itself was almost sheer, and the men were having trouble enough finding their footing without being knocked off balance by falling debris or having stones slide out from under their moccasined feet. Mackenzie climbed to the top of the bank to view the route ahead and call out warnings and instructions to his men – uselessly, as it turned out, because the thunderous roar of wildly-tumbling waters completely drowned out his voice. In the course of the next two miles, the rapids were so bad that the party – including its leader – had to portage the canoe's contents no less than five times.

For three exhausting days they fought the Peace River, until "it began to be muttered on all sides that there was no alternative but to return." Yet for all their labors, when Mackenzie climbed part way up the side of the canyon, he could see nothing but a steady succession of rapids and falls. Since the river route was impassable, Mackenzie finally set off across country on what turned out to be a seven-mile hike.

At daybreak on the fourth morning, they began what even the habitually casual author of the *Voyages from Montreal* admits was an "extraordinary journey."

> The men began without delay to cut a road up the mountain, and as the trees were but of small growth, I ordered them to fell these which they found convenient in such a manner that they might fall parallel with the road [being made], but at the same time not separate them entirely from the stumps, so that they might form a kind of railing on either side . . . the whole party proceeded with no small degree of apprehension to fetch the canoe . . . we advanced with it up the mountain, having the lines [of rope] doubled and fastened successively as we went on to the stumps; while a man at the end of it hauled it round a tree, holding it on and shifting it as we proceeded; so that we may be said, with strict truth, to have warped the canoe up the mountain.

Although he does not say so at this particular point in his book, Mackenzie had to keep his men going by an admixture of his own calm behavior and generous handouts of liquor.

The Peace River Canyon was an experience that shattered their confidence. Hitherto they had mastered all the waters of the Northwest. Until that day, they had never encountered the fury and menace of a mountain river in flood, particularly one which, as even Mackenzie admitted, was "one white sheet of foaming water." After the traverse around the Canyon, Mackenzie kept his terrified, near-mutinous men moving westward for weeks on his own reserves of willpower and courage. The only thing *he* feared was failure to complete his task of discovery.

At the western end of the portage, the voyageurs spent half a day trimming fresh tree branches with which to pole the canoe upriver and ease the labor of paddling against the powerful current. Here, the pragmatic Mackenzie left a visiting card for local Indians as a friendly gesture and an invitation to trade: he erected a pole, and attached to it a packet containing a knife, a steel flint, beads, and a few other minor articles.

As they journeyed westward some days later, snow-capped peaks towered on either side of them – the Rocky Mountains. It was almost the end of May and the sun shone clearly, but the dry cold of the region was so penetrating that even the hardy canoemen complained of its numbing effect on their hands as they worked the poles. The weather was so bitter that they had to wear their blanket coats day and night. To keep them good-humored, Mackenzie ordered an issue of rum, usually at the end of each day's labor. By May 29 a whole keg had been consumed. The weather became so cold on the 31st that he called a halt at nine in the morning with the intention of kindling a fire to thaw out his half-frozen crew – "a very uncommon circumstance at this season." However, he thought of a quicker solution: "a small quantity of rum served as an adequate substitute."

They passed through the Rocky Mountains and came to what in later years was called Finlay Forks, where the Finlay River, racing down the Rocky Mountain Trench from the north, joined the Parsnip River from the south to form the Peace River. West of the Forks was a chain of mountains "running south and north as far as the eye could reach." Which waterway were they to take? The Finlay led north and looked easier, and the men made it plain that it was their choice. Mackenzie agreed "as it appeared to me to be the most likely to bring us nearest to the part where I wished to fall on the Pacific Ocean." But he recalled that an aged Beaver warrior he had met earlier that year at Fort Fork had advised him

> . . . not on any account to follow it [the Finlay], as it was soon lost in various branches among the mountains, and that there was no great river that ran in any direction near it; but by following the latter [the Parsnip], we should arrive at a carrying-place to another large river, that did not exceed a day's march, where the inhabitants build houses and live upon islands. . . .

He ordered the steersmen to proceed up the swifter, narrower Parsnip.

It was a crucial decision. The Finlay would have led the party to where the headwaters of the Liard, Skeena, and Stikine rivers interlock in a maze of streams; in all likelihood they would have lost their way completely. Very sensibly, Mackenzie heeded the old man's warning, although as a result he had to endure days of bitter complaint from his men and, worse still, occasional doubts as to the wisdom of his choice.

Paddling occasionally, but more often poling against the strong current, the canoemen moved their craft over the wild waters of the Parsnip. The work was utterly exhausting because the farther upriver they advanced, the stronger the current became. Poling was often the only method of making progress because the Parsnip was overflowing with spring runoff, and the banks of the river were so heavily lined with trees that it was impossible to use a tow line. Sometimes the men could only maintain a forward motion of sorts by pulling on the branches of partly submerged trees.

On June 5, during part of which Mackenzie had left his party for several hours to view the country from a nearby height, they reported to him on his return that

> the canoe had been broken, and that they had this day experienced much greater toil and hardships than on any former

> occasion. I thought it prudent to affect a belief of every respresentation that they made, and even to comfort each of them with a consolatory dram [of rum]. . . .

But Mackenzie had to admit that the canoe had been reduced to "little better than a wreck" and needed extensive repairs. To add to their miseries, the weather on the west side of the Rocky Mountains was much milder and gnats and mosquitoes "appeared in swarms to torment us."

Eight days upstream from Finlay Forks, he became desperately anxious to find the carrying-place that the aged Beaver had said would take him to "the larger river . . . where the inhabitants build houses and live upon islands." Although Mackenzie didn't know it, he had already missed the mouth of a waterway (the Pack) that would have taken him across to that river – probably because of his habit of indulging in a short doze in the canoe every now and again. Late that night, over a supper of boiled wild parsnips and pemmican, Mackenzie fretted about the location of the carrying-place. Had he missed it during one of his naps? Had he actually seen it and mistaken it for one of the numerous island channels on the river? Did it exist at all?

On June 9th, guided by a local Sekani, Mackenzie and his men left the main stream of the Parsnip behind, traversed a two-mile-long body of water (Arctic Lake), and then made a short portage. After the portage, for the first time they were, in their leader's words, "going with the stream." They were on water that would find its way to the Pacific Ocean.

Some weeks and several hundred miles later, Mackenzie led his jittery crew out onto salt water at the mouth of the Bella Coola River in a borrowed, Indian canoe. (Their own had slowly disintegrated under the pounding blows of the wild mountain waters they'd traversed.) So Mackenzie had not found a navigable route through the confusing jumble of mountain ranges between the Rockies and the Pacific coast. But his characteristically stubborn refusal to accept defeat from man and nature alike made his failure seem a triumph.

It is for this reason that he's in all the school books. Yet Mackenzie has rarely been given credit for something else he tried – and failed – to do. That particular aim is incorporated in the *Voyages from Montreal.* He published the book not just to draw attention to himself but to prod the British government into acquiring territory in western North America before the newly-established United States of America did so. He offered the general reader a travel book in the form of an explorer's logbook – but he also offered those in high places an economic treatise on the fur trade and policies for a mighty expansion of it under the British flag. The book's preliminary discourse, entitled "A General History of the Fur Trade from Canada to the North-West," occupies *one quarter* of the entire book.

The two summers Mackenzie spent exploring were essentially minor episodes in a thirty-three-year career, much of which was devoted to reorganizing and unifying the Canadian fur trade. In the process of trying to bring about this reorganization and unification, Mackenzie qualifies as a very early father, perhaps a grandfather, of the federation of provinces called Canada. Why? Because barely twelve months after his body was laid to rest in 1820, the British government was induced, principally by the lobbying of Montreal and London fur merchants, to arrange a merger of the North West Company and the Hudson's Bay Company under the name of the latter, and grant it a monopoly of trade throughout the entire western interior and on the Pacific coast. This union, formally signed on March 26, 1821, vindicated the logic of Mackenzie's commercial views. Unable to overcome the Hudson's Bay Company's great advantage of cheaper, direct access to the continental interior, and thus to the Pacific, the Nor'Westers were forced by years of steadily rising costs to accept terms of union with the Bay organization. As for Mackenzie's political reasons for a unified, expanded fur trade, under the leadership of another Highland Scot, George Simpson, this new, stronger Hudson's Bay Company went on to challenge the Russians and the Americans on the Pacific coast and succeeded in establishing a British claim to a large part of the coastal Northwest. This claim had far-reaching consequences. In 1846, when the British and American governments negotiated a settlement of the boundary beyond the Rocky Mountains, the United States secured what was known as the

Oregon Territory. But Great Britain retained those regions immediately north of the 49th parallel that later became known as British Columbia. Without access to the Pacific Ocean, Canada would probably have been unable to resist absorption into the United States. Without an outlet on the Pacific, it is extremely doubtful if there would ever have been a Canada.

* * *

The history of the fur trade is the story of a slow but steady movement from one over-trapped region to the next untouched one westward. Thus it was fellow traders who followed Mackenzie's track up the Peace River. A Canadian, and some years later a Scot, ventured onto the headwaters of the Peace. It is the first of these two men whose name appears on the map today in the words "Finlay River," the waterway that joins with the Parsnip to form the Peace River.

John Finlay was the son of Christiana Youel of Montreal and James Finlay, who was one of the first British trader-merchants to hustle his way into the former French fur trade operated out of Montreal. The senior Finlay seems to have been the more colorful character and certainly the more successful, becoming the founder of Finlay & Gregory, a highly enterprising firm of Montreal fur merchants (who were, among other things, the employers of a young fur trader named Alexander Mackenzie). If the son had any hopes of inheriting the family business, these were killed early on. James Finlay sold out to John Gregory, his English co-partner, and another trader-merchant, and retire (to become, of all things, an inspector of chimneys in Montreal). Gregory, McLeod as the firm came to be called, was soon forced to save its corporate life by joining that hardnosed, highly successful business monopoly known to Canadian history as the North West Company. So, at the age of fifteen, John Finlay signed up as a Nor'Wester clerk (apprentice trader) the year that Mackenzie was on his way to the Arctic. Three years later, Finlay joined him on the Peace River as assistant trader in charge of another fur post. The records of the time are irritatingly scanty, but two particular incidents in John's life have been documented.

In the fall of 1792, Mackenzie was canoeing west on the Peace River to winter as far upstream as possible before making a dash overland to Pacific tidewater. On the grey, chilly morning of October 19th, he came to an abandoned North West Company post wreathed in smoke and flame. Mackenzie and his voyageurs hastily landed and were able to save some of the smaller houses, cursing the men who had stopped off at the post overnight and carelessly left a fire burning in the main building. Mackenzie says that, later that day, he caught up with the miscreants – a trading party led by eighteen-year-old John Finlay.* (Doubtless, the latter got well and truly chewed out by the sales manager of the Athabasca Department.) Five years later, Finlay was still working in the Peace River region, and it was either in 1797 or the following year that he spearheaded trading operations west of the Rocky Mountains (as yet another Nor'Wester, Simon Fraser, would do in 1805-1808). Finlay journeyed through the beautiful prairie countryside alongside the river until he reached the mouth of a stream (Tea Creek) a few miles west of the modern town of Fort St. John. Here he built the first trading post in what is now the Province of British Columbia. Or he may have constructed Rocky Mountain House after exploring farther upriver. (If Finlay kept a diary of his trading activities and explorations, it has not been found. And North West Company documents are so terse, so thoroughly businesslike that many of the details of its operations have been lost forever.) There is, however, in one trader's journal, an unmistakable reference to Finlay's charting the lower reaches of what is described as "Finlay's river."† And this would not have been exploring for exploring's sake. A company run by Scots was nothing if not money-minded, and Mackenzie had reported of the headwaters of the Peace River that "in no part of the North-West did I see so much beaver-work."

* *Voyages from Montreal* . . ., London and Edinburgh, 1801.

† Rich, E. E., ed., A *Journal of A Voyage From Rocky Mountain Portage in Peace River To the Sources of Finlays Branch And North West Ward In Summer 1824* [*BY SAMUEL BLACK*], London, The Hudson's Bay Record Society, 1955, p. 197.

The supposed discoverer of the Mackenzie's farthest source soon fades out of history, but the identity of the real explorer of "Finlay's River" remained unrecognized for the next 120 years. That's a long story. It's enough to say here that, in the 1920s, it became something of a hobby with J. N. Wallace, a Dominion Land Surveyor, to investigate the activities of the wintering partners on the Peace River and in the Athabasca country. The book he ultimately published on the subject suggests that Wallace had all the doggedness and patience of a detective. It was in the course of comparing old North West and Hudson Bay company records with various historical accounts of the fur trade, checking and cross-checking diaries and journals, and analyzing surveying notes left by several trader-explorers that Wallace uncovered a startling fact overlooked by several professional historians. The explorer of the Finlay River was, beyond all doubt, a Scot named Samuel Black.*

Black was born out of wedlock and, some say, really was a bastard. During the gang warfare waged between the North West and Hudson's Bay companies for some years prior to their amalgamation in 1821, Black earned a reputation as a Nor'Wester "bully." Enormous in stature, he was described by a contemporary as "A Don Quixote in appearance . . . rawboned and lanthorn-jawed, yet strong, vigorous and active. . . ." (It says something for Black that, when a young clerk, on some occasion or other, he didn't hesitate to "have words" with as domineering a personality as Alexander Mackenzie.) A painfully honest man who voiced his opinions in slow, solemn speech, he was accorded great respect by his trade colleagues, whether Nor'Westers or Bay men. In fact, Black managed to intimidate his opponents, terrorize their customers, and antagonize HBC executives to such an extent that he, Peter Skene Ogden, and Cuthbert Grant were the only Nor'Westers excluded from the union of 1821. In all fairness to Black and his fun-loving friend, Ogden, they had been instructed to make pests of themselves and did so with great thoroughness. Neither actually used physical force, strong men though they were. But they did

* Wallace, J. N., "The Explorer of Finlay River in 1824", *Canadian Historical Review*, Vol. IX, 1928, pp. 25-31.

have a great time ripping the bark off HBC post roofs and stuffing it down chimneys, beating on walls in the middle of the night, stealing their rivals' fishing nets, scaring off game, and taking occasional potshots at an HBC flag or weather vane.

Black's career as a trader came to a temporary end in 1821 with the merger of the two companies into a new, stronger Hudson's Bay Company. George Simpson, a cocky young HBC executive and the only opponent Black was never able to intimidate, had a hand in his exclusion. But Simpson – another certified bastard, this time a West Highland one – recognized, in his own words, a "very cool, resolute to desperation" character when he saw one and remembered him when he had need of a man "equal to the cutting of a throat with perfect deliberation." (Simpson's many judgments of his subordinates have to be taken with a large grain of salt. He was somewhat cynical even when in the best of humors. As the saying goes, a man is known by his friends. And Black's closest friend for many, many years has been described by a contemporary as "Peter Ogden, the humorous, honest, eccentric, law-defying, the terror of Indians and the delight of all gay fellows."*) By 1823, Simpson had convinced the London board of governors that Black's "zeal" and his well-known "exertions" were needed by the Company. Within two days of arriving back in Canada, Black was given orders by Simpson to explore beyond the Rocky Mountains "to the source of Finlay's Branch and the North West Ward." Simpson was just beginning a long and very successful career as a hard-driving corporate leader who was forever lecturing his subordinates about "Oeconomy," as he termed it, and the need to "extend the Trade to Countries hitherto unexplored." There was no doubt in his mind that Black was exactly the man to lead – or drive – a party of voyageurs into the wild mountain country of Finlay's River and "introduce us to a new and valuable branch of Trade."

For his part, Black promptly accepted the task. It was a job and, tacitly, a forgiveness of past sins. But he had a much more personal reason for taking on this tough assignment. He had read and reread the reports of such explorers as

* Cox, Ross, *Adventures on the Columbia River*. . ., London, 1831.

James Cook, Alexander Mackenzie, and Simon Fraser and hungered to join that highly exclusive club. Black's account of his exploration* has many a revealing entry that suggests his dearest wish was to make a name for himself as an explorer: as he wrote on one occasion in the flickering light of the evening campfire, "I wanted to do something more & discover other Rivers & Lands. . . ." It was in desperate search of just such fulfilment that Black sternly fought down the naturally quick-tempered side of his character and sought by patient example, joking remark, and a strangely steady compassion to lead his party of voyageurs through the green, moss-carpeted forests of the Rocky Mountain Trench and up into the dripping-wet, storm-lashed uplands of the Cassiar Mountains.

Wild, indeed, was the river Black had been ordered to explore. The outstanding, utterly extraordinary, feature of this part of Canada is the Rocky Mountain Trench, an immense trough anywhere from 2 to 10 miles wide that separates the Canadian Rockies from the other western ranges. Down the northernmost 300 miles of the Trench races the frothing, impatient water of the lower Finlay. It twists and turns in frenzied efforts to find a quick way past massive piles of driftwood, all the time gradually ridding itself of a surface load of dead or dying trees and an underwater one of glacial gravel and silt. This section of the Finlay created enormous problems for Black's party. There were islands made up of monstrous piles of timber torn out of the forested river banks; and innumerable bends where tangled wood debris lay just beneath the river's raging surface, ready to rip open a birchbark canoe and expose its occupants to the hazards of rock and rapid. Then they had to pole, portage, and pull their way up the Big Bend of the Finlay, a great, canyon-strewn turn the waters make as they drop down from the Stikine Mountains into the Trench.

Thereafter, the going got steeper and steeper. Each day was an exhausting battle, hauling the twenty-five-foot canoe up a fast, clear, ice-cold waterway that sapped their

---

* Rich, E. E., ed., *A Journal of A Voyage From Rocky Mountain Portage in Peace River To the Sources of Finlays Branch And North West Ward In Summer 1824*, London, The Hudson's Bay Record Society, 1955.

strength and froze their limbs. Finally, after circling round rapids and falls, Black and his men reached the haven of the calm, fir-rimmed waters of Thutadé Lake. As if nature itself wanted to bestow a reward on Black and his dog-tired men, it was here one evening that they had a markedly better-than-average supper; a bellyful of tender steak cut from several hard-swimming but luckless, caribou. Yet, in order to do so, the explorers had had to paddle, portage, pole and pull their birchbark craft up the equivalent of a 2,000-foot hill of water – and this in the course of one of the coldest, wettest summers ever known in northwestern Canada. (Black remarks that as late as June 3, ice formed on the gunwales and paddles.)

Although Black went 100 miles beyond Thutadé, on foot this time, he was dogged by a terrible sense of failure. The farther into the northwest he penetrated, the wilder the countryside became and the wetter the weather. He met few natives and saw little evidence of beaver. As George Simpson summed things up in a letter the next year to the Company's London governors, "the country was so bare of the means of subsistence that they could not proceed nor attempt to pass the Winter where they were as starvation would have been the consequence . . . they had no alternative but to retrace their steps to Peace River . . . having made no discovery of importance."

The astute, all-knowing Simpson was both right and wrong. Black had not discovered "a new and valuable branch of Trade." As his boss brusquely reported, "there are a few Beaver, but occupying such inaccessible places . . . they are not sufficiently numerous to defray the expenses of an Establishment . . ." But there *had been* another, important discovery, of no consequence whatever to the businesslike Simpson and of little comfort to poor, disconsolate Black: Thutadé Lake is, geographically speaking, the farthest source of the Mackenzie River.

For almost a hundred years, Black and his men were the only persons to explore Mackenzie waters to their ultimate source in a creek at the western end of Thutadé Lake. (Even today, only a handful of people can boast the same achievement.) The list of trader-explorers who mapped a country "with too much geography" is a long one, and goes back three centuries to Jacques Cartier. To that list must be added the names – typically a mixture of Indian, French, Métis,* and Scots – of a stubborn adventurer and his faithful companions: Samuel Black of the parish of Pitsligo, Aberdeenshire, Scotland; his second-in-command, Donald Manson, of the town of Thurso, Caithness, Scotland; Joseph La Guarde, of French and Iroquois origin, and, like his leader, a former Nor'Wester; no less than three other ex-Nor'Westers – Joseph Courneyer of Sorel, Lower Canada (Quebec), Antoine Perreault, who may have been a Métis, and Jean Baptiste Tarrangeau, who seems to have been a Métis, and likely born and bred in the Red River district; Baptiste La Prise, a Chipewyan hunter-interpreter, and his wife; and a native guide, only identified by Black as "the old Slave," and his family.

* * *

The Peace River did attract the occasional tourist. Perhaps the most famous of these was Lieutenant William Butler of Her Britannic Majesty's 69th Regiment of Foot, whose romantic fascination with what he termed the "Wild North Land" induced him to travel upriver in the dead of winter in the course of a rather roundabout journey from Fort Garry, Manitoba, to Victoria, B.C. (His passionate interest in the Northwest did not, however, prevent Butler from wryly observing that "the Roman nose could not have withstood an Arctic winter, hence the limits of the Roman Empire.") And, as most of us dutifully learned in elementary school, it was the river that led the fur-trading Simon Fraser and his successors into a savagely beautiful land so reminiscent of Scotland that the only possible name for the region was New Caledonia. But it was that other great Mackenzie tributary that became a much-travelled highway to and from the Pacific. Much the southerly of the two, the Athabasca was more easily reached from the prairies.

Exactly who first followed the upper waters of the Athabasca River into the awesome, confusingly look-alike

* The French term for a person of Indian and European blood. Most native people pronounce it "may-tee" but some prefer "may-tiss."

ranges of the Rocky Mountains will never be known. Perhaps it was a Kootenay, Assiniboine, or Blackfeet hunter, desperate for moose, bear, or buffalo with which to feed and clothe his family and their numerous relatives. Or it may have been a wandering group of Beaver Indians, whose tribe was driven farther and farther into the Rockies in the eighteenth century by the Cree, those mercenary middlemen of the HBC fur trade. This discovery may even have been made by bands of Nipissing (Ojibwa) or Iroquois, whom generations of Canadians have learned to associate with the Great Lakes, the St. Lawrence River, and the story of New France. Oddly enough, these "immigrant" Indians were working as hunter-trappers in the Rockies between the time Mackenzie initiated the push to the Pacific in 1793 and the follow-up journey by another Nor'Wester, Simon Fraser, thirteen years later. Encouraged by the North West Company to go west, most of these Nipissings and Iroquois spread out in the forested foothills country on the east flank of the Rocky Mountains between the Athabasca and Peace Rivers. A few seem to have wandered up the Athabasca and explored the Jasper area. One of them, known to history simply as "Thomas the Iroquois," was the man without whom David Thompson, yet another Nor'Wester, might never have found his way through the Rocky Mountains and over the Athabasca Pass. That pass made possible, for several decades, the first trans-Canada highway.

David Thompson, Welsh by birth and a Hudson's Bay Company man by early training, was only a fur trader out of sheer necessity. It was a job that fed and clothed him. But it did take him to hundreds of out-of-the-way places, all of which enabled him to indulge a fantastic passion for surveying and mapping. The result was that, in the course of twenty-four years of fur trading, Thompson managed to explore, plot, and chart almost two million square miles of what is now western Canada. He was as familiar with the long course of the Peace River as he was with that of another great western highway, the Saskatchewan River. He could find his way through the watery, insect-infested country around Hudson Bay as easily as he navigated the mountain-born waters of far western rivers. It was towards the close of this long career of travel and exploration that David Thompson made a westward journey that fulfilled Alexander Mackenzie's dream of a water route to and from the Pacific Ocean.

In the period 1807-1810, Thompson was the sales manager of what the North West Company called its Columbia Department – all the lands west and south of the Rocky Mountains, in effect, parts of today's southern British Columbia and much of northern Oregon, Idaho, and Montana. During these years, he was in the habit of crossing and recrossing the Rockies via a pass halfway between the modern resort towns of Banff and Jasper. It was not the best of crossing places: the western side was heavily timbered and the ground choked with deadfall and thick brush. Nor had it aided Thompson in finding a navigable river that led to the Pacific. He still had to journey around his Department on horseback, on foot, and by canoe. But at least this pass got him through the massive, saw-toothed barrier of the Rockies. However, in the fall of 1810, he and his men found their way barred by angry Piegan warriors, who were, in their own words, the "frontier tribe of the Blackfoot nation." They guarded the eastern approaches to the Rocky Mountains against intruders. And they had a special reason for halting Thompson. He was selling goods and guns directly to the Kootenay and several other tribes living west of the mountains. In so doing, he was ruining an extremely profitable trade that the Blackfoot had been engaged in for some years as middlemen between the North West Company (and, for that matter, the Hudson's Bay Company) and these transmontane tribes. So there was no way these Piegan scouts were going to let Thompson's party go on up the North Saskatchewan River, into the mountains, and over the Howse Pass.* Luckily for Thompson, in a chance encounter with a Nor'Wester colleague, he learned that there was

* Just as the Finlay River is misnamed, so is the Howse Pass. Called after an HBC trader who shadowed Thompson to find out where he had discovered new customers, it seems to have been used by Kootenay venturing into Piegan territory in search of buffalo. The first man known to have blazed a trail through this pass was Thompson's interpreter and scout, "Jacko" Finlay, half-brother of the fur-trading John Finlay.

another crossing place near the headwaters of the Athabasca River, one that was said to have been successfully used by Nipissing fur hunters.

Thompson and his men set off across country, hacking a trail through a heavily-timbered foothills region. Although they loaded all their goods and belongings on pack horses, it took them the better part of a month to cover the two hundred miles to the Athabasca. Guided by Thomas the Iroquois, seven weeks of trudging upriver in windy, sub-zero temperatures brought them to the entrance of today's Jasper National Park. A few miles later Thomas led them to a small, windowless shack.

Rested and refreshed, Thompson and his voyageurs plodded on up the ice-covered river. At this point, most of the horses were so spent that the trade goods had to be transported on dog sleds hastily constructed out of logs. The weather remained well below zero as the party slogged on into ever deepening snow in which the dogs floundered and the sleds stuck time after time. Where the tributary of the Miette joins the Athabasca (just south of the modern resort of Jasper), Thompson had no option but to lighten the sleds by caching some of his goods. It was in this neighbourhood, too, that he turned loose those horses that had not been slaughtered for food; some of his men now had to shoulder the packs these horses had carried.

Thirteen men and eight dog sleds labored on south up the wide, wooded valley of the Athabasca until Thomas the Iroquois halted and stared hard at a particular snowcapped peak on the west side of the river. Satisfied that he recognized what voyageurs would later call the "Mountain of the Grand Crossing" (and we call Mount Edith Cavell), he pointed the way up a frozen waterway (the Whirlpool River) that came from the southwest out of a long, mist-shrouded valley. At a word from Thompson, the little party slogged on and trekked painfully but steadily up to the 5,000-foot level.

On the morning of January 10th, 1811, Thompson and his men were almost clear of the timberline. He sensed that they were very close to the continental divide, the height of land that twists and winds its way through the mountains from Alaska to Mexico and determines which waters flow east and which west. Sure enough, that night, a clear and brilliant one, Thompson walked a little way from the campfire and found that he was in an immense gap in the mountains. Far below, he saw a ribbon of ice that wound towards him and then swung around a big bend and moved away towards the southwest. It was the Columbia. Thompson had discovered the navigable waterway to the Pacific that Mackenzie (and Simon Fraser) had sought in vain. In effect, Thompson had found the Pacific-coast equivalent of the Saskatchewan River, the great highway of the prairies. Now, thanks to the happy circumstances of a pass that could be crossed fairly easily and could be reached via the Athabasca River – whose waters are Mackenzie waters – the continent could at last be crossed all the way to and from the Pacific.

A day or two later, as his men stumbled and plunged through the deep snows and tall timber of the western slopes of the Rocky Mountains, David Thompson was unaware that he was blazing a trail many would follow. Largely because of the continual threat of pillage, if not of violence, from Piegans, the crossing he had just made became a regular route of the Nor'Westers, and, later, of the men from the Bay. Once time and money had been spent clearing a trail through the heavy timber on either side of the Athabasca Pass, traders and supplies poured over this crossing place and down to the big, hairpin bend of the Columbia River that became known as Canoe Encampment. From here, they journeyed into the "Inland Empire", today's southern British Columbia and the state of Washington. From its territories, Nor'Westers would haul thousands of pelts over the pass until, in 1821, they were absorbed into the Hudson's Bay Company. Then, for many a decade, HBC canoe brigades poled and paddled the Athabasca, freighting packs of goods one way and bales of furs the other.

* * *

Men from the Bay were by no means the only transcontinental travellers on Athabasca waters. In dribs and drabs, various non-commercial characters turned up on the river. As usual, Scots were in the vanguard, this time scholars as desperate for knowledge as their commercial brethren were for profits.

The first of these early tourists was Thomas Drummond. A naturalist who was supposed to accompany Sir John Franklin's second expedition down the Mackenzie to the Arctic, Drummond's fascination with the plant and animal life of the prairies completely sidetracked him. Instead of accompanying the other members of the expedition north into the Mackenzie basin, he wandered off west along the Saskatchewan River on his own, avidly listing specimens as fast as he could collect them. Eventually, Drummond turned up at the HBC's Edmonton House, where he persuaded the factor (trader in charge) to loan him a horse and a guide so that he could set off for the Rocky Mountains and investigate its fauna and flora. So in 1825-1826 Drummond spent a blissfully happy year roaming about in the Jasper region, sometimes in company with Iroquois hunter-guides, sometimes on his own and living on whatever fish or game he could find. Cold, hunger, and solitude didn't discourage his passion for listing and classifying various species of plants, birds, insects, and quadrupeds. Nor did an encounter with a female grizzly, which offered to do battle with him. Drummond, searching for moss specimens, had quite unknowingly come between this mother and her two cubs. When his gun misfired because the powder was damp, he calmly reversed the musket and with the confidence born of sublime ignorance prepared, in his words, to "disable" the bear by giving it a hefty whack on the head. Luckily, the sudden appearance of an east-bound HBC brigade changed the odds, and the bear escaped "unhurt . . . and apparently intimidated by the number of the party." What finally did discourage Drummond – who was just about to wander over the mountains to the Columbia River – was a curt message from his commanding officer to rejoin the expedition, now getting ready to return to England.

A year later, another Scots scientist, whose name is immortalized by the king of the western forests, the mighty Douglas fir, arrived in the same region on his way home. He had been, in effect, a one-man collecting expedition for Britain's Royal Horticultural Society and had been roaming the North Pacific coast for three years. Travelling in company with an east-bound HBC brigade and carrying his precious botanical specimens and notes in a tin box, the tiny but tough David Douglas steadfastly refused all offers of a horse. He insisted on tramping all the way over the Athabasca Pass and along the edges of the Athabasca River – just in case he missed any specimens in the, to him, brand-new country east of the continental divide. Anyway, a horse was a bit of a nuisance if you suddenly wanted to dash into a grove of trees on the chance of finding a new species of fir or fern or scramble up a snow field to dig for the bulbs of glacier lilies.

Mackenzie headwaters made transcontinental travel possible but not pleasurable. The valley of the Athabasca was a traumatic shock to tourists who'd just ambled across grassy, sun-drenched prairies conveniently stocked with high-protein meals in the shape of buffalo and deer and which even provided occasional stop-over facilities in the form of an HBC post. Once across the height of land that separates waters flowing to the Arctic from those draining down to Hudson Bay, those travelling west found themselves in a drastically different landscape, one dominated by gloomy forests, evil-smelling swamps, turbulent streams, and raging rivers. And for almost a thousand miles between Fort Edmonton on the upper waters of the Saskatchewan and Fort Vancouver on the lower reaches of the Columbia, the only roofed dwelling was a draughty, ramshackle, windowless log hut. Indeed, the only impressive thing about Jasper House was its name.

Travellers' descriptions of this section of their journey are uniformly horrifying. In 1859-60, James Carnegie, Earl of Southesk, arrived on the scene in the role of big-game hunter, heading for Jasper House under the guidance of several Métis and Iroquois hunters. But the noble earl and his men never did get all the way into the mountains: the countryside due west of Fort Edmonton was so exhausting that the party eventually veered south into more open territory along the eastern flanks of the Rockies. Day after day they rode along an old Indian hunting trail through "dense woods, chiefly of poplar brush with a few firs, and often through deep morasses filled with fallen timber." There was barely room for a loaded horse to pass between the trees. "By the end of the day my knees were one mass of bruises from cannoning off the fir stems." Southesk was profoundly grateful for a present that the factor of Fort Edmonton had given him, a leather hunting shirt, for "no woollen clothes but the stoutest can stand against these horrible thickets." Thickets, as Southesk politely called the tangled, thorny brush, were so dense that when one of the pack horses strayed it was never found, although a guide tried very hard to follow the animal's tracks. Even more exhausting were the extensive stretches of muskeg that had to be crossed. Every now and then, a horse would sink up to its belly in the slime and mud or, in a fit of panic, thrash around and flop over on its side, to sink slowly up to its neck. Then came the nightmare job of wading into the brackish muck, stripping the terrified animal of its pack load, and hauling it free with lengths of rope. Luckily for Southesk, his own mount had a phlegmatic temperament. It was "expert at crossing deep ravines and cared nothing for muskegs, however deep and bad; even when sinking in a swamp he would take the opportunity to snatch a bite of grass if his nose got near enough the surface." Scrambling over fallen timber on steep hillsides, Southesk's horse would sometimes "while drawing himself over a log, stop half-way and begin eating any tempting mouthful that happened to be handy."*

The next tourists, unlike Southesk, travelled alongside the Athabasca itself. They endured the same basic difficulties and had much trouble finding a place where they could cross waters enormously swollen by the meltwater of a winter of heavy snowfalls. In 1863, yet another British aristocrat, William Fitzwilliam, Viscount Milton, his physician-cum-general manager, Dr. Walter Butler Cheadle, and an impoverished Irish schoolmaster, professional scrounger, and general incompetent called Eugene Francis O'Byrne, or O'Beirne, set out from Fort Edmonton in order to accommodate Lord Milton's bizarre wish to "visit" the Cariboo goldfields of central British Columbia. (O'Byrne had wheedled his way into the company of Milton and Cheadle by claiming to be a fellow graduate of Cambridge University.) This

* Southesk, Earl of, *Saskatchewan and the Rocky Mountains. A Diary and Narrative of Travel, Sport, and Adventure, during a Journey through the Hudson's Bay Company's Territories, in 1859 and 1860*, Edinburgh, 1875.

curious trio, plus a couple of Métis guides, took the long-established fur-trade route north of Fort Edmonton to the Athabasca and then turned westward. With its muskeg, burnt trees, dense underbrush, and thick timber, the trail was an eye-opener: "no one but a Hudson's Bay voyageur would dream of taking horses into such a region," the good doctor noted.*

The party's daily routine was a regular comedy of errors. At every stream, his lordship stopped to pan for gold. O'Byrne, who wouldn't go anywhere near a horse if he could help it, nonetheless insisted on giving totally unwanted advice on how to load and unload the pack animals. The guides, of course, often scolded Cheadle for O'Byrne's calculated laziness. And there were some mornings when his lordship felt so fragile and unable to face the day that he refused to rise before noon.

A continuing problem was their inability to find a safe ford across the swift waters of the Athabasca. Yet, they had to cross it in order to reach another trail that wound a torturous way over the continental divide via the Yellowhead Pass. And the only way to get across the river was by laboriously felling large pines and making a raft. (As usual, O'Byrne waited until he thought everything was done and then came fussily forward to offer his help.) "Heavy" was Cheadle's description of the tossing, churning waters that barred their way; even the horses would have to be ferried over. So it seemed a good idea to sit down for a while and toast success. As a matter of fact, they finished off part of a barrel of rum, which probably accounts for the safe arrival of men and animals on the far bank and explains Cheadle's cheerful description of the crossing as being done "famously."

Messrs. Milton, Cheadle, and O'Byrne were very fortunate. They encountered the Athabasca in summer; several travellers experienced the river at its winter worst. One was the painter, Paul Kane, who recorded the pains and perils of this season in the course of a round-trip journey from Toronto to the Pacific.

---

* Cheadle, Walter B., *Cheadle's Journal of Trip across Canada 1862-1863*, with Introduction and Notes by A. G. Doughty and Gustave Lanctot, Ottawa, 1931.

It was the mountain section that exhausted Kane most. On the outward trip, November gales and snowstorms combined to make him cold and wet every day. After enduring twenty-three days of poling and pushing their boats up a "shallow, icy river," the fur brigade and its solitary passenger arrived at Jasper House in the middle of a howling blizzard. In fact, at the height of the storm, the boat Kane was in was blown clean out of the water and hurled, he says, fifteen feet up the river bank; he had to transfer to a horse and four times ford a river "sometimes coming over the saddle" before reaching the safety of Jasper House. However, he was thrilled to reach the top of the Athabasca Pass and there see

> a small lake . . . this forms the head waters of one branch of the Columbia River on the west side of the mountains, and of the Athabasca on the east side. It is about three-quarters of a mile of circumference, and is remarkable as giving rise to two such mighty rivers; the waters of the one empyting into the Pacific Ocean, and of the other into the Arctic Sea.*

When returning along the same section of the route exactly a year later, the weather was much worse. Once back alongside the Athabasca, "It became so cold that we could no longer sit on the horses, but were obliged to dismount and drive them on before us. My beard, the growth of nearly two years, gave me much trouble, as it became heavy with ice from the freezing of my breath; even my nostrils became stopped up, and I was forced to breathe through my mouth." Fortunately, the HBC brigade came across an Indian lodge in which Kane had an opportunity to thaw himself, after which he rode on to Jasper House in comparative comfort. There, he forgot his troubles over "a good piece of mountain sheep, which is really delicious, even when not seasoned by such hardships as we had undergone."

Despite the bitter weather, Kane preferred not to spend a dreary winter in the wretched accommodation Jasper House offered and set off with two guides for Fort Assiniboine, from

---

* Kane, Paul, *Wanderings of an Artist among the Indians of North America from Canada to Vancouver's Island and Oregon through the Hudson's Bay Company's Territory and back again*, London, 1859.

which he could easily reach the greater comfort of Fort Edmonton on the North Saskatchewan. A sled carried Kane's baggage and precious sketches, plus a few provisions, because "we trusted to our guns to provide more on the way," and Kane blundered along on snowshoes that were exactly his own height: five feet eleven inches. The three men had about three hundred and fifty miles to go on the part-frozen surface of "the most monotonous river that ever I have met with in my travels. Nothing but point after point appearing, all thickly covered with pine, any extensive view being entirely out of the question."*

As the trio plodded across the frozen lakes the Athabasca forms just east of the modern resort of Jasper, Kane was amazed to meet two Indians walking across the ice in their bare feet. Sitting down with them for a smoke, he discovered that it was an old native habit to take off both snowshoes and moccasins when walking on ice or frozen snow: anyone accustomed to this method of walking claimed to suffer less from it than from the sweat that accumulated inside moccasins in the course of quick travel, froze, and then cracked into small pieces and lacerated the soles of the feet. If Kane didn't believe their story then, he did about ten days later when he found inside his stockings slivers of frozen perspiration as much as an eighth of an inch thick. When some of these slivers worked their way down inside his moccasins, Kane felt as if he was walking barefooted on sharp gravel.

Day after day, their progress became slower and slower. In many places where the river flowed rapidly in summer, the current had jammed chunks of ice one on top of another to form what voyageurs called *bourdigneaux*, hillocks and ridges ten to fourteen feet high. At the edges of the river, deceptively thin, often snow-covered, layers of ice had formed as a result of a temporary damming of the river's flow; a traveller was well advised to poke a stick ahead of him to avoid crashing through into the waters still running below. Yet the trio dared not leave the river for the even more treacherous rock-strewn ground and tangled forest alongside its banks.

With all the climbing, sliding, and falling, it wasn't long

---

* *Ibid.*

before Kane began to suffer the agonies of *mal de racquet.*

This complaint attacks those who are unaccustomed to the use of snow shoes, if they walk far on them at first. It is felt at the instep. I do not know how to convey an idea of the intense pain, except by saying that it feels as if the bones were broken, and the rough edges were grinding against each other at every motion . . . the men, seeing me suffer so much . . . recommended me to scarify my insteps, and kindly offered to perform the operation, which is done with a sharp gun-flint; but I was afraid of the frost getting into the wounds, and refused, although I had every confidence in their knowledge of what would be the best remedy in a case like mine.*

To make matters worse, they hadn't been able to shoot a single animal en route, and what provisions they had were rapidly diminishing. Twelve days out from Jasper House, all their food was gone, and they began to look hungrily at the sled dogs. Kane and his men finally decided that even these two, elderly, skinny animals would not make much of a meal for three hungry men. In any case if the dogs were eaten, the three would only further weaken themselves by having to take turns at hauling the sled.

On the fifteenth day, hunger woke them early. Long after the brief northern daylight had faded, the men staggered on downriver, using the reflected light of the aurora borealis to guide them over or around the glistening humps of ice strewn on the surface of the Athabasca. Finally, they saw ahead of them spirals of smoke curling up into the blue-black bowl of the night sky and realized they had at last reached the security of Fort Assiniboine. There, Paul Kane ate himself into insensibility – although he didn't do so until he had eased his conscience by feeding raw fish to the sled dogs.

The memory of that feast hung over me, even in dreams, for many a day afterwards . . . seated on a pile of buffalo skins before a good fire, I commenced the most luxurious repast of which it had ever been my fate to partake. I had no brandy, spirits, nor wine, neither had I tea or coffee – nothing but water to drink. I had no Harvey's sauce, or catsup, or butter, or bread,

* *Ibid.*

> or potatoes, or any other vegetable. I had nothing but fish; no variety, save that some were broiled on the hot coals and some were boiled. But I had been suffering for days from intense cold, and I now had rest; I had been starving, and I now had food . . . in the morning, about five o'clock, I commenced again and made another hearty meal; and then how happy I was when I lay down and slept again, instead of clambering over the rugged bourdigneaux!*

* * *

Every last one of these adventurous tourists – and several more who might have been mentioned – would not have been able to tour the Northwest without the advice, transportation services, and generous hospitality of the Hudson's Bay Company. Indeed, had it wanted to do so, the Company could have refused to allow any of these people to travel west of the Great Lakes because the Company more or less owned what is now western and northern Canada. The Company was, in fact, a state within a state. By a charter granted by the English Crown in 1670, the Hudson's Bay Company could establish its own army and navy, administer justice (both civil and criminal) and "execute justice accordingly," and even declare war "on any prince or people not Christians." In addition, all officers of the English navy, army, and civil government were commanded to give the Company aid in its endeavors to engage in trade in those territories of North America draining down to Hudson Bay. In return, the Crown simply demanded that, whenever a member of the royal family entered these territories, he or she be given "two Elkes and two Black Beavers."

The history of the Hudson's Bay Company is older than the history of Canada. And, of course, in free-enterprise terms, the Company is still going strong, although its major activities are now the operation of department stores, the mining, smelting, and refining of ores, and the expansion of a flourishing gas-and-oil industry. However, no business organization survives for three centuries without having to endure a fair number of ups and downs. And it was in the vast drainage basin of the Mackenzie River that the Company fought its bitterest battle for survival.

---

* *Ibid.*

Chapter 3

A Battleground for Furs

The fur trade existed for several centuries because of a persistent fashion in men's hats.

It was in the late sixteenth century that the beaver hat first became a common headgear among wealthy Europeans. Then the fashion really caught on, and for the next two hundred and fifty years there was a steadily-increasing demand, a demand that ultimately became something of a mania. Whether royal or aristocratic, clerical or lay, the male of the species rarely appeared in public without wearing a broad-brimmed, glossy-smooth, beaver hat. It might be plumed, bejewelled, braided, or embroidered; tall, short, large, small, or tricorne in shape according to the fashion of the moment. To flaunt such a hat was to proclaim one's membership in high society. Like a sports car or a mink coat, it was a status symbol. Of course, a beaver hat was also a highly-visible indication of wealth. That great seventeenth-century society gossip and diarist, Samuel Pepys, records paying £4 for one in a day and age when a much-sought-after architect like Sir Christopher Wren earned all of £200 a year. Indeed, it was not uncommon for a man to bequeath his beaver hat in his will.

This demand for an article of clothing made from the shorn, pressed, and steamed pelt of a large-sized member of the rodent family gradually cleaned out Europe as a source of supply. However, just about the time that beaver was becoming an exceedingly rare species in the Old World, unlimited numbers of that animal were being discovered in the northern half of the New World. Although the luxurious sheen and hard-wearing quality of the beaver pelt dominated the trade from the start, other furs were also utilized.

Second-rate hats, together with cloak trimmings and ladies' muffs, could be made from otter, muskrat, ermine, fox, lynx, mink, fisher, and marten skins. Thus was born the rivalry between French entrepreneurs operating out of Quebec City, Trois Rivières, and Montreal and the English employees of the Hudson's Bay Company.

With one or two wartime exceptions when there were raids and counter-raids that disrupted British trading activities, this rivalry never amounted to very much. After all, hundreds of miles of rock, lake, and forest separated the Great Lakes-St. Lawrence region from the shores of Hudson and James bays. However, various Frenchmen managed to find a canoe route as far west as the Assiniboine and Saskatchewan rivers, waterways that led them onto the prairies. This enabled them to intercept customers on their way to HBC fur posts – all of which were situated alongside salt water. In 1755, it was reported to the Company's London board of governors that one of its "servants" had actually witnessed the strength of competition from trade rivals. Anthony Henday, or Hendry, who had explored all the way from the Bay almost to the Rocky Mountains themselves was halfway home again when he discovered a "French Trading House" on the North Saskatchewan River. Here, his Indian companions "received from the Master ten Gallons of Brandy half adulterated with water" and eagerly traded "nothing but what were prime Winter furs. . . . It is surprising to observe what an influence the French have over the Natives; I am certain he hath got about 1000 of the richest Skins."* About a week later, much the same thing happened at another post farther downstream. This time, Henday noted "the French talked Several [Indian] Languages to perfection: they have the advantage of us in every shape; and if they had Brazile tobacco, which they have not, would entirely cut off our trade." As Henday sadly reported, the French allowed all the least desirable pelts to be taken on down to the Bay – a journey of several hundred, very difficult, miles.

---

* Burpee, Lawrence J., ed., "York Factory to the Blackfeet Country: The Journal of Anthony Hendry, 1754-1755," Royal Society of Canada, *Transactions*, Section II, 1907.

HBC management paid little attention to Henday's report. Yet he had observed all the elements of successful fur trading: save the customer time and trouble by meeting him on his home territory, soften him up with friendly talk and handouts of liquor, and then cream off the thickest, glossiest furs. When, four years later, New France was conquered by the deadly musketry of Wolfe's regiments, these were exactly the rules followed by a number of men who took over the French fur trade in its entirety – language, transport methods and routes, and trading customs. Individually at first, and then collectively, they caused the Hudson's Bay Company tremendous trade losses for the better part of half a century. In so doing, they expanded the trade all the way to Arctic and Pacific shores. And this development is, in part, the story of Alexander Mackenzie's career as a fur trader, executive, and lobbyist.

* * *

The first reliable information about Mackenzie's career is that, in 1779, at the age of fourteen or fifteen, he began to earn his living as a bookkeeper in the counting-house of Finlay, Gregory & Company, a firm of Montreal fur merchants. A few years later, he was working as a trader in the continental interior. In the Preface to the *Voyages from Montreal*, Mackenzie tells us why he chose this particular activity.

> I was led, at an early period of life, by commercial views to the country north-west of Lake Superior in North America, and being endowed by nature with an inquisitive mind and enterprising spirit, possessing also a constitution and frame of body equal to the most arduous undertakings, and being familiar with toilsome exertions in the prosecution of mercantile pursuits . . .

In a guarded manner and in rather formal prose, Mackenzie admits to seeking his fortune. And this could be done in the fur trade, provided that you were healthy, hardy, and lucky enough not be be drowned – and also if you did not starve or freeze to death.

Unlike the exclusively British-administered, British-manned Hudson's Bay Company, the adventurers who competed for furs in the forests and plains to the north and west of the Great Lakes, a vast stretch of territory that their voyageurs called *le pays d'en haut* (the high or upper country), were of several nationalities. But they shared a common characteristic: a restlessness that drove them to adopt a wilderness way of life. As was often remarked by officials and clergy alike in the days of the French régime, "A trader is a rebel at heart." The fur trade seemed to satisfy their need for total freedom and provided them with opportunities to make a great deal of money. A few were French Canadian: Maurice Blondeau, who wintered with prairie Indians as early as 1752, was trading at several plains posts some years before Finlay first canoed up the Saskatchewan. Some of these "Pedlars from Montreal," as the men from the Bay contemptuously called them, were from the American colonies, notably a smooth speculator and handsome huckster called Alexander Henry. Even more hailed from Europe, from England, and, in particular, from the Highlands of Scotland. Mackenzie was destined to meet most of these men and to work with some of them. A handful became notable figures in the history of the trade, specifically as "Nor'Westers" - members of various combinations of Montreal fur merchants and western-based "wintering partners" known to history as the North West Company.

While a fur trader was essentially an adventurous individual, he had to be tough-fibred and self-sufficient. He had chosen a way of life that exposed him regularly to danger and discomfort - in particular, the threat of starvation and the prison-like confinement imposed by the long, bitter winters of the *pays d'en haut*. If a trader was located on the prairies, he could hunt buffalo, deer, and game, or simply purchase meat, venison, and fowl from local Indians. Every plains post had a huge *glacière* or icehouse packed each fall with tons of animals carcasses that included buffalo, deer, black bear, swans, geese, and ducks. Yet while buffalo steaks and roasts were satisfying and sustaining, a plains trader was also careful to secure large supplies of the fatty flesh from the animal's hump and back. A mixture of fat and lean meat prevented a man from contracting the sickness that results from a low-fat diet. But in the vast forest belt north of the prairies trading posts had to be built alongside lakes in order to catch fish in great quantities, because there was little to eat but fish all winter long. Of all those caught – whitefish, sturgeon, trout, pike, carp, sucker, bass, and bream, most of which are rich in natural oils – the huge sturgeon of the region was most prized. David Thompson, who spent several winters in the heavily forested country along the Churchill River, described it as the "freshwater hog," adding that "whatever is not required for the day is frozen and laid by in a hoard, and with all care is seldom more than enough for the winter."

Thompson did not exaggerate. The threat of famine always menaced the northern posts. There are many records of death from starvation or of traders and voyageurs reduced to roasting bear hides, or eating broths made from beaver skins or from a rock lichen known as *tripe de roche*. In his book, *Travels and Adventures in Canada and the Indian Territories Between the Years 1760 and 1776* (Boston, 1809), Alexander Henry says that the recipe for the latter was simply a matter of boiling the lichen "into a mucilage, as thick as the white of an egg." It was unappetizing but filling, although eating tripe de roche produced unpleasant side effects. John Long, another trader, remarks that this glutinous mess caused "violent pains in the bowels, and frequently occasions of flux. . . . When the disorder does not terminate in a flux, it occasions violent vomiting, and sometimes spitting of blood, with acute spasms of the bowels."

In fact, almost all a trader's difficulties were the result of living and working in a northern climate. He spent the brief summer season when the rivers were open transporting furs several hundred miles to a central depot – usually Grand Portage on the north shore of Lake Superior, or Michilimackinac at the tip of the Michigan peninsula – and then racing back to his post with packs of trade goods before ice put an end to all navigation. For much of the year he had to endure frigid weather with French-speaking voyageurs for company. Next to hunger, his great enemy was boredom: there was little for him to do. Daniel Harmon, a Vermonter who spent twenty years in the *pays d'en haut*, noted in his diary that "leisure moments" accounted for "nearly nine-tenths" of his time. A trader's duties consisted of little more than extending credit to Indians who occasionally came in to replace a broken musket or buy supplies of ammunition, or entertaining hunters and their families who journeyed to the post to barter furs. If anything, his hardest work was the trading session – dispensing liquor and cheap bribes of combs, mirrors, and beads and then haggling for hours, sometimes days, in order to secure as many furs as possible for the least outlay in goods. Small wonder that a trader often trudged a hundred miles or more through weather that ranged from 30 to 60 degrees below zero to visit another, equally remote, post. Here, he could at least speak English with a compatriot and enjoy with him a few days of reminiscence and gossip, and not a few hours of drunken forgetfulness. These were welcome breaks in an otherwise monotonous, brutish existence that was made even more frustrating by the voyageurs' natural love of the wilderness and their casual, easy relationships with Indians, which few traders could, or even tried to, emulate. It is clear from their journals and diaries that most of these early traders despised their customers, and many of them treated their Indian wives or mistresses at best as slaves, at worst as animals.

Among these Pedlars from Montreal working in the *pays d'en haut* was one whose activities in the Northwest would greatly influence Mackenzie's career and, as a result, quicken the expansion of the fur trade into the drainage basin of the Mackenzie River. He was a born wanderer from Connecticut called Peter Pond.

Contemporary writers seem to agree that Pond was a loner, and a short-tempered one at that, although he always seemed to get on well with Indians and in his early days gained quite a reputation as a peacemaker. He appears to have been a very confident, aggressive man, which is hardly surprising because he was one of a tough breed in a tough business, men who thrived on cut-throat competition and regularly engaged in price wars. Since the great demand in Europe for beaver hats showed no signs of decreasing, in the Great Lakes region in the late 1760s and in the *pays d'en haut* in the mid-1770s the tempo of competition quickened.

The Pedlars became steadily greedier in their dealings. Their only concern was to grab pelts by the hundreds, if not by the thousands. It didn't matter how many Indians were cheated or robbed outright in the process. It was of no consequence whether a trader used liquor, guile, or threats – or all three –provided that he could send canoe-loads of beaver pelts back each summer to his sponsors in Montreal.* And if a man consistently outsold his immediate rivals, he was highly rated, even by those whom he had outsmarted. Pond became a leading member of this unholy fraternity.

After spending a few years trading on upper Mississippi waters, Pond decided to try his luck north of the Great Lakes, where more – and better – furs were evidently to be had for the taking. But competition from an aroused Hudson's Bay Company, which was at long last sending its servants up the Saskatchewan to build posts on the north branch of the river, cut into Pond's trade. During two successive winters the returns of his labors were much smaller. Having heard vague rumors about fine furs being bought north of the Saskatchewan on the Churchill River, he decided to go there and get a share of this business.

As it happened, Pond ended up several hundred miles farther north and west of where he had originally intended to go. His plans were changed by the accident of meeting a group of Pedlars, who had sensibly decided to try the experiment of joining forces instead of competing among themselves. They pooled their surplus goods and provisions and invited Pond to act as their representative and trade for furs in a region called Athabasca, which was said to be northwest of the headwaters of the Churchill. It was on this river three years before that Alexander Henry, in company with two fellow traders, had intercepted Indians coming down from Athabasca on their way to Hudson Bay. These Chipewyans were acting as middlemen between the Hudson's Bay Company and other Indians living far to the north and west

* Some of the merchants who provided the traders with goods and canoes were located in Detroit, Albany, and Quebec City. But most of them ultimately settled in Montreal, which became the fur capital of North America. Even the most successful of all the American fur merchants, John Jacob Astor, used to make annual journeys to Montreal to purchase pelts.

where winters are eight or nine months long and animal furs correspondingly thick and glossy. Henry and his companions had been amazed by the magnificent quality and quantity of the Athabasca pelts obtained from the Chipewyans. Pond's sponsors were quick to appreciate the importance of Henry's discovery and they wisely picked Pond to advance their interests.

The journey along the Churchill River was a nightmare. There were dozens of portages caused by wildly plunging cascades and rapids; in numerous places the river races violently through narrow, twisting, rocky channels; and its placid stretches of water often concealed masses of tree debris that were close enough to the surface to tear gaping holes in birchbark canoes. Once as far west on the Churchill as Lake Ile-à-la-Crosse (in northwestern Saskatchewan), Pond's searchings took him northwest to a twelve-mile portage across the height of land that separates waters flowing to Hudson Bay from those flowing to the Arctic Ocean. Once across the punishing Methye Portage and embarked on the Clearwater, a tributary of the Athabasca River, Pond found himself in a landscape completely different from those along the Churchill or Saskatchewan rivers. He had entered a subarctic region which, with its hundreds of lakes and thousands of square miles of muskeg, is more water than land. It has been aptly remarked of Athabasca that "the country is four-fifths drowned and when not frozen is half-hidden by mosquitoes and black flies." Pond was now at 59°N, almost three thousand miles from Montreal. But as his canoes moved through this desolate, uninviting country, he was looking at the richest fur region the trade would ever know.

What sets Pond apart from his commercial brethren is not just his ability to outsell most of them, but a rather unexpected trait: he proved to be a talented geographer and map maker. Sometime in the 1780s, he became preoccupied with charting the rivers, mountains, and ultimate boundaries of the regions beyond his post, which was just a few miles upriver from Lake Athabasca. He may have been stimulated by Indian reports that Russian fur traders had established a trading centre somewhere on the Pacific Coast of North

America. In any case, Pond was extremely curious about the Pacific Ocean. How did one reach it overland? What routes led to it? And how did the Arctic Ocean fit into the topography of the Northwest? Was it true, as Indians told him, that there was a mighty river north of Lakes Athabasca and Great Slave that ran down to a frozen sea? Pond spent many a frigid Athabasca night trying to piece together the jumbled mixture of what he knew and what he was told – by Indians, by his traders, and by their voyageurs – much of which was vague and conflicting in nature. Eventually, he managed to work out for himself one key fact of the geography of the Northwest. The Athabasca River, which flowed past his post, and the even larger Peace River that his informants said "descended from the Stony or Rocky Mountains" formed the forked tail of a mighty water system that flowed far to the north.

Unfortunately for Pond, and even more unfortunately for his employers – plus Alexander Mackenzie, who later served under Pond in Athabasca – he went on from there to mess up the geography of the Northwest. In his desperation to work out a canoe route to the Pacific, Pond, after reading the text and examining the maps of Captain James Cook's published account of his efforts in 1776-1779 to find the Pacific gateway of the fabled Northwest Passage, took a wrong turning. He deceived himself into thinking that Cook's description of a waterway emptying itself on the south coast of Alaska was the western end of another waterway that, Indians told him, flowed out of Great Slave Lake. One result of this was Mackenzie's attempt, in 1789, to reach the Pacific by venturing north of Athabasca. It was only after the crushing disappointment of ending up at the Arctic that Mackenzie, very wisely, adopted Pond's original thinking and went west up the Peace River – and thus reached his goal.

But, while Peter Pond had unknowingly figured out a route to the Pacific, he hadn't solved the logistics and finances of maintaining a fur-trade operation based thousands of miles east in Montreal. That's where the genius of Alexander Mackenzie comes to the fore. Thanks to his time in the counting-house of Finlay & Gregory adding up bills, plus his experience of actually trading with customers, he was beginning to work out the answer to the problem of keeping sufficient control over expenditures to ensure a profit.

At all times, the business of bartering for pelts was one that came close to being a form of gambling. Much money was risked and much money was gained, but the chances of bankruptcy were high. A fur company could be ruined by several factors over which it had little or no control: a poor trading season for various inexplicable natural reasons; outbreaks of hostility among native groups; epidemics of European diseases, to which Indians were very susceptible; adverse fall weather that forced early wintering in over-trapped areas, or particularly violent spring rains and runoff that greatly increased the normal chances of canoes foundering and men and cargoes being lost. But what came to concern Mackenzie most were two interlocking factors. Slowly but surely during the 1770s and 1780s, the Montreal fur trade had been reaching farther and farther into the Northwest. Thus more and more time was needed to transport goods into the interior and to bring furs out. Yet the whole operation was based on a credit system that became costlier the longer it took to pay off creditors. He explains this last difficulty very clearly in the "General History" section of *Voyages From Montreal*.

> The agents [fur merchants] are obliged to order the necessary goods from England in the month of October, eighteen months before they can leave Montreal, that is, they are not shipped from London until the spring following, when they arrive in Canada in the summer. In the course of the following winter they are made up into such articles as are required for the savages; they are then packed into parcels [pièces] of ninety-pounds weight each, but cannot be sent from Montreal until the May following so that they do not get to market until the ensuing winter, when they are exchanged for furs, which come to Montreal the next fall and from thence are shipped, chiefly to London, where they are not sold or paid for before the succeeding spring, or even as late as June; which is forty-two months after the goods were ordered in Canada, thirty-six after they had been shipped from England, and twenty-four after they had been forwarded from Montreal. Thus, the merchant,

> allowing that he has twelve months' credit, does not receive a return to pay for these goods and the necessary expenses attending them, which is about equal to the value of the goods themselves, till two years after they are considered as cash, which makes this a very heavy business.

Longer supply lines to and from the *pays d'en haut* and rising costs were not the only difficulties. Competition was becoming keener with the gradual emergence of a particular combination of Montreal fur merchants and their traders. Known as the North West Company, it threatened to take over the Athabasca region, the richest source of pelts in the *pays d'en haut*. As Mackenzie records, thanks to co-operation between some traders, Pond was able to reach Athabasca with enough goods to buy a small fortune in furs. And the rewards of co-operation were plain for all to see. It was, as Alexander Henry observed, "beneficial to the merchants, but not directly so to the Indians, who, having no other place to resort to nearer than Hudson's Bay, or Cumberland House [for many years, the only HBC post in the western interior], paid greater prices than if a competition had existed." This conclusion was also reached by the merchants operating certain Montreal fur firms, particularly when they saw and handled the thick, heavy, lustrous pelts that came from Athabasca. Each was struck by the same thought: with superb furs such as these, a man could make a fortune several times over. The result was a prolonged effort by these men to make Athabasca (and much of the rest of the *pays d'en haut*) their exclusive territory. They established co-partnerships with each other and also with a number of "wintering partners," each firm's senior traders in the *pays d'en haut*, and used the handy title of the North West Company to describe their common business activities.

The history of the North West Company, which probably began sometime in the middle or late 1770s, is complex not least because its membership was often a changing, usually expanding, one. The name alone is confusing because the organization itself was composed of a number of individual fur companies. However, the North West Company was never at any time in its existence a company in the normal sense of a legally constituted corporation accountable for its actions, financing, etc. It was simply a trading name used by a loose association of men who worked in the same line of business and agreed to turn it into a very profitable livelihood by creating what amounted to a monopoly. Despite its loose nature, the Company was an immensely successful combination of experienced wintering partners – Nor'Westers as they proudly called themselves – and a number of well-financed Montreal fur companies. No one has described these entrepreneurs better than the Canadian historian, W. Stewart Wallace:

> The story of the Nor'Westers, though not without its darker pages, is a brilliant chapter in the history of Canada. No braver or more picturesque band of adventurers ever put it to the touch, to gain or lose it all. Some of them were French-Canadian traders and voyageurs, the sons of those who had followed La Vérendrye to the rivers and prairies of the West in the dying days of the French régime. Others were American frontiersmen who had served their apprenticeship in the fur-trade in the valleys of the Ohio and the Mississippi. Most of them were Scottish Highlanders, the sons of those who had come to Canada in Wolfe's army or as United Empire Loyalists in the American Revolution. The number of them who were connected with that gallant regiment, the 78th or Fraser's Highlanders, is remarkable; and it is no less remarkable that of the numerous Frasers, McTavishes, and McGillivrays, who played such an important part in the history of the North West Company, nearly all came from Lord Lovat's estates [in the north of Scotland]. The names of the North West Company partners sound like a roll-call of the clans at Culloden. These men were hardy, courageous, shrewd, and proud. They spent a good part of their lives travelling incredible distances in birch-bark canoes, shooting rapids, or navigating inland seas. They were wrecked and drowned. They suffered hunger and starvation. They were robbed and murdered by the Indians, and sometimes by one another. They fell victims of smallpox, syphilis, and rum. Yet they conquered half a contintent . . .*

* Wallace, W. S., ed., *Documents Relating to the North West Company*, Toronto, The Champlain Society, 1934.

With few exceptions, Nor'Westers were men who displayed unrelenting opposition to anyone who was not of their number. They acted as if they *owned* the *pays d'en haut*. Whether competing with traders working for other Montreal companies or with the men from the Bay, they often used Mafia-like tactics - threatening violence, hijacking their competitors' supplies of goods, and bribing, bullying, or intoxicating Indians into becoming their customers. Many Nor'Westers in effect controlled groups of Indians by making them dependent upon handouts of liquor. If these tactics didn't work, their colleagues in Montreal sometimes neutralized competition by bringing rivals into the Company as fellow partners - which is what happened to Gregory, McLeod, Mackenzie's employers - and giving them a share of the profits. These colleagues also handled the Company's public relations. Every fur merchant had to obtain annual government licences in order to trade in the *pays d'en haut*. Thus the Montreal members were always careful to give officials in Quebec the impression that their wintering associates were an admirable group of trader-explorers who were busy extending the limits of British influence and territory in North America and described themselves as merchants devoted to fostering the flow of furs and goods between Quebec and London. Ruthless, powerful, but outwardly respectable, the North West Company was one of the earliest examples of "Big Business" in North America.

In his "General History", Mackenzie gives us an inside view of the Company's very simple but very successful financial structure, which had been cleverly organized to offer each member a strong incentive to work as hard as he could. Mackenzie's account explains exactly why the Nor'Westers were such fierce competitors: some partners put up the money to finance the Company's operations, other partners formed the sales force, but all the partners received a share of the profits. Even the lowly clerks (apprentice traders) knew that, in time, they would become partners and therefore profit sharers. As Mackenzie points out, the Company

was no more than an association of commercial men, agreeing

among themselves to carry on the fur trade. . . . It consisted of twenty shares, unequally divided among the persons concerned. Of these, a certain proportion was held by the people who managed the business in Canada, and were styled agents for the Company. Their duty was to import the necessary goods from England, store them at their own expense at Montreal, get them made up into the articles suited to the trade, pack and forward them, and supply the cash that might be wanting for the outfits [clerks' and voyageurs' wages, and the purchase of food supplies], for which they received, independent of the profit on their shares, a commission. . . . The remaining shares were held by the proprietors [traders] who were obliged to winter and manage the business of the concern with the Indians. . . . Some of them, from their long service and influence, held double shares and were allowed to retire from the business at any period of the existing concern with one of these shares, naming any young man in the Company's service to succeed him in the other [by buying the share]. . . . Thus all the young men succeeded in succession to the character and advantages of partners. . . . This regular and equitable mode of providing for the clerks of the Company excited a spirit of emulation in the discharge of their various duties. . . . Indeed, without such a spirit, such a trade could not have become so extended and advantageous, as it has been and now is.

The clerks were paid a salary that varied with length of service and was anywhere from £100 to £400 annually, and they were also supplied with food and clothing. The partners, of course, received their money when the year's annual profit was calculated and then divided by the number of shares. In the period 1790-1795, the annual Company profit averaged £72,000 and was in the £90-100,000 bracket for each of the last four years of the century – and this in a day and age when an income of £1,500-£2,000 a year was considered a comfortable living. Even as late as 1802, when fur-trade rivalries recurred and many of the best fur regions had been practically trapped out, a North West Company share was worth close to £3,300.

Of course, the Montreal fur trade was utterly dependent upon its French-Canadian canoemen. Only tough, tireless voyageurs had the strength and stamina to take literally hundreds of tons of goods thousands of miles into the continental interior. Whether they were *mangeurs du lard* (pork eaters) manning a *canot de maître* between Montreal and Grand Portage, or *hommes du nord* (northmen), the élite canoemen who went into the *pays d'en haut* in the smaller, lighter *canot du nord*, they were the indispensable workhorses of the trade. Their basic skill was to take a fragile, heavily laden, birchbark canoe safely through the seething white waters of a *sault* or, more laboriously and equally dangerously, to "line" or haul their craft all the way past it. Their endurance was legendary. It was commonplace for crews to paddle forty strokes to the minute for fifteen hours a day with only two meal breaks in that time and a few short rest spells of ten or fifteen minutes each. They averaged 4 to 6 miles an hour in calm waters and, weather permitting, could keep up this pace for weeks on end. Head winds or rough weather brought the risk of swamping and put an end to progress, sometimes for several days. Such was a voyageur's natural impatience with any delay that he described his inactivity as *dégradé*, and "degraded" quickly became part of the vocabulary of English-speaking travellers on the canoe routes of the *pays d'en haut*. The voyageur was nothing if not vain and never happier than when displaying his strength and dexterity for all to see; many a traveller records his inability to keep up with a canoeman carrying two ninety-pound pièces over a stony portage trail at a steady dogtrot.

To the *engagé*, the salaried voyageur of the fur companies, wind, rain, snow, and sleet were all part of the job. The pay, which varied between £15 and £50 a year, was out of all proportion to his hours of work and the conditions under which he performed them. Yet he thought himself fortunate to receive money and a yearly issue of a blanket, a shirt, a pair of trousers, and a few pounds of twist tobacco. (He received twice as much if he was a northman.) When travelling, he greatly relished the plainest of daily rations: a mush of corn, peas, and pork or bacon fat on the two-month run from Montreal to Grand Portage or, in the *pays d'en haut*, pemmican in raw chunks or boiled to make what was called "rubaboo."

The voyageur followed a rough, ill-paid, but, above all,

dangerous occupation. Canada's lakes are notorious for their sudden summer squalls and storms, which spring up in a matter of minutes and can pound ships, let alone canoes, completely under. On the rivers of the *pays d'en haut* there were few large saults without tall wooden crosses clustered on the banks to mark the last resting place of an entire crew whose craft had foundered amid jagged rocks and savagely tumbling waters. And alongside stretches of certain prairie waterways a number of crosses bore mute testimony to the effectiveness of Sioux lances or Blackfoot arrows launched during an ambush. Even the stresses and strains of portaging made hernia a prevalent complaint and not infrequently a cause of death.

In fact, death was always around the next bend of the river or somewhere out on the waters of the lake ahead. But to the breed of man who chose to become a voyageur, his gaiety and pride were such that the presence of death gave added spice to life. The fur trade was an escape from the dull, grinding poverty of rural Quebec and offered him opportunities to see and experience a wider, freer world. A gregarious fellow, he sustained himself with the company of his comrades – and the occasional Indian girl – and his desire to travel to as many places as his employers saw fit to send him. He loved "the forest and the white water, the shadow and the silence, the evening fire, the stories and the singing and a high heart." And if he was picked to become a northman, he regarded this as the finest of compliments and the passport to an even better and more adventurous life. Its reckless satisfactions are clear from the perhaps exaggerated but obviously happy reminiscences of an aged homme du nord, who preferred to live out even his remaining years in the *pays d'en haut*.

> I have now been forty-two years in this country. For twenty-four I was a light canoeman; I required but little sleep, but sometimes got less than required. No portage was too long for me; all portages were alike. My end of the canoe never touched the ground till I saw the end of it. Fifty songs a day were nothing to me. I could carry, walk, and sing with any man I ever saw. During that period I saved the lives of ten Bourgeois [wintering partners], and was always the favorite, because when others

stopped to carry at a bad spot, and lost time, I pushed on – over rapids, over cascades, over chutes; all were the same to me. No water, no weather ever stopped the paddle or the song. I have had twelve wives in the country; and was once possessed of fifty horses and six running dogs, trimmed in the first style. I was then like a Bourgeois, rich and happy; no Bourgeois had better-dressed wives than I; no Indian chief finer horses; no white man better harnessed or swifter dogs. I beat all Indians at the race, and no white man ever passed me in the chase. I wanted for nothing; and I spent all my earnings in the enjoyment of pleasure. Five hundred pounds, twice told, have passed through my hands; although I now have not a spare shirt to my back, nor a penny to buy one. Yet, were I young, I should glory in commencing the same career again. I would spend another half century in the same way. There is no life so happy as a voyageur's life; none so independent; no place where a man enjoys so much variety and freedom as in the Indian country. Huzza! huzza! pour le pays sauvage.*

This, then, was the superbly-organized business machine that sent, first, Alexander Mackenzie, then Simon Fraser, and finally David Thompson, to find a navigable water route from Montreal to the Pacific and back. However, the Nor'Westers never did succeed in duplicating their rivals' enormous advantage of a cheap supply route by sea that brought trade goods all the way into the heart of the continent at Hudson Bay – as opposed to a two-month canoe journey to get from Montreal to the same latitude. And, of course, the same HBC merchantmen took, on average, a mere four weeks to transport cargoes of furs to London, England – as opposed to a four- or five-month haul to Montreal from Athabasca or the Pacific coast, plus a six- or seven-week voyage from Montreal to London. In other words, it cost the Nor'Westers more money to market their furs than it did the HBC.

Despite this severe financial drawback, for several decades the North West Company's aggressive ability to bribe or bully Indians into becoming their customers brought the Hudson's Bay Company close to financial collapse. Indeed,

---

* Ross, Alexander, *The Fur Hunters of the Far West: A Narrative of Adventures in the Oregon and Rocky Mountains*, 2 vols., London, 1855.

after years of yielding a 10% return, HBC dividends dropped to 8% and, by 1800, had sunk to 4%. But the real struggle for domination of the Athabasca country was about to break out and was initiated by Alexander Mackenzie.

In 1799, ten years after his voyage to the Arctic and six after his journey to the Pacific, he was working as a fur-trade executive in Montreal. "Nor'West" Mackenzie was one of that city's better-known figures, something of a hero to many of the younger wintering partners in the North West Company, and financially secure. But he was happy with neither his position nor his prospects in McTavish, Frobisher & Co., which had become the virtual directorate of the North West Company. He was the fifth of five co-partners and had little say in determining overall Nor'Wester policy. A long-cherished plan he had to expand, regulate, and control the entire continental fur trade – including the operations of American rivals – had not been advanced one whit. Mackenzie seems to have excited his friend and fellow executive, William McGillivray, with his notions of establishing "one common interest" to supervise the trade and to conduct all freighting of goods through Hudson Bay. Mackenzie had even impressed these same ideas on Colonel John Graves Simcoe, Lieutenant Governor of Upper Canada, as early as 1794, the year in which he had left the *pays d'en haut*. On his way to Montreal, Mackenzie had stopped off at Newark (Niagara-on-the-Lake) to pay his respects to the governor, and they enjoyed a long chat together. Finding him "as intelligent as he is adventurous," Simcoe listened attentively to Mackenzie's proposals, which included the establishment by the British government of two posts on the Pacific coast as a claim to sovereignty, and incorporated these in a long report on western problems that he sent to London. In Simcoe's account is the first intimation of the project that Mackenzie tried to promote over and over again in later years.

> . . . the most practicable Rout to the Northwest was thro' the territories of the Hudson's Bay Company; that by the Rout from Great Britain all the Navigation from Montreal thro' the chain of Lakes & their immense Communication to the most distant

part of the interior Country & its consequent Carriage would be saved.*

Time and again, Mackenzie tried to persuade his fellow partners of the logic of his project, but failed to convince them. Finally, he decided to quit the North West Company and appeal to a much larger audience. In mid-December 1801 the *Voyages from Montreal* was published simultaneously in London and Edinburgh, a large, quarto volume of 550 pages illustrated by maps and containing an engraving of the author from the portrait by Sir Thomas Lawrence. The book became a bestseller. Aided by generally favorable reviews and made popular by serialization in magazines, within two years of publication it had reappeared in several editions, including three in the United States. The book was later translated into French, German, and Russian. Barely a month after the publication of the *Voyages*, the business-like Mackenzie submitted to the British government a detailed plan of fur-trade expansion (which was more or less ignored by Whitehall) and then returned to Montreal to lead his own fur company against both Nor'Westers and HBC men. Like the North West Company, his firm was an association of several firms, officially registered as Sir Alexander Mackenzie & Company, although the more common historical reference to it is the XY Company.

The loyalty one Highlander expects of another runs deep, and Mackenzie's defection from Nor'Wester ranks had been neither forgotten nor forgiven. When, in 1804, the basic problem of punishing costs obliged Nor'Wester and XY men to merge and try to form a stronger North West Company, Mackenzie was allowed the status of a partner – but he was denied any role in management. After a brief, curiously uncharacteristic, career as the representative of the electors of Huntingdon County in the provincial legislature of Lower Canada, he returned to England to become a fur-trade lobbyist. Introducing himself to the British government this time as a North West Company partner, Mackenzie requested a charter for his colleagues that would give them the

* From an appendix to the report that accompanied Simcoe's dispatch No. 36 of 11 September, 1974, Public Record Office, London.

same monopoly rights on the Pacific coast that the Hudson's Bay Company enjoyed in the interior of the continent. Nothing came of this and similar, later efforts at lobbying. Because he had good reason to distrust the slow progress of government towards a point of ultimate action, in 1808 Mackenzie began to buy Hudson's Bay Company stock. (This, thanks to the Nor'Westers, had sunk from £100 to £60 a share.) His object was to purchase control of HBC operations and secure at least Hudson-Bay transit rights for the Nor'Westers, if not also exclusive trading rights in Athabasca. Ironically, Mackenzie was aided in his objective by another, much wealthier Scot, Thomas Douglas, fifth earl of Selkirk, who had his own reasons for seeking power to make HBC policy. Lord Selkirk's interest was, in his own words, "at the western extremity of Canada, upon the Waters which fall into Lake Winnipeck [*sic*]," where "with a moderate exertion of industry" colonists could be sure of "a comfortable subsistence." He dreamed of founding a settlement at Red River.

By 1810 Mackenzie, Selkirk, and a group of associates had invested several thousand pounds in HBC stock. But by this time the former had discovered his ally's motive – the establishment of a settlement in the Nor'Westers' Red River Department – and Mackenzie ended up opposing Selkirk's plan. With his usual astuteness, Mackenzie sensed trouble: if farming spread all over a region where buffalo, the source of huge Nor'Wester supplies of pemmican, were accustomed to roam freely, the great herds would abandon the Red River region. This could create immense difficulties, if not hardship, for men serving in the *pays d'en haut*.

But, by 1811, Selkirk had not only persuaded the Governor and Committee of the Hudson's Bay Company to accept his colonization scheme, but had gained total personal control of it. Mackenzie could do little thereafter but harass Selkirk's lieutenants in various minor ways as they organized the dispatch of the first group of settlers. It was from this set of circumstances that there developed a bitter, bloody, battle between Selkirk and the Nor'Westers. Hereafter, Sir Alexander Mackenzie gradually fades out of history, to die nine years later, an honored and respected figure, but a neglected, powerless man. However, before he died peacefully on the way to his country estate in the north of Scotland, other men had died violently in the far-off watershed of the Mackenzie River.

* * *

As early as the winter of 1812-1813, a rumor raced along the Peace River from one Nor'Wester post to the next that the Hudson's Bay Company was going to invade the region. Months passed, and nothing happened. Finally, in the fall of 1815, John Clarke of the HBC led a force of almost one hundred men from the Bay into Athabasca.

Clarke's employer was a company that had traditionally regarded the fur trade of North America as a personal possession, yet had done very little to combat a combination of Scots and French Canadians with a highly-developed regard for free-enterprise methods of doing business. The HBC had lagged far behind its rivals in establishing trade outlets within the *pays d'en haut*. Shortly after the turn of the century, it tried to maintain a post at Lake Athabasca, but had been so thoroughly intimidated by Nor'Wester threats of violence that the HBC withdrew and contented itself with competing side-by-side with its rivals' forts along the Saskatchewan River. One hundred and forty years after its founding, the Company had no post west of the Rocky Mountains or north of the prairies. By 1815, however, things had changed.

The driving force behind Clarke's expedition was not so much the HBC itself as it was the indomitable will of one particular shareholder. Thomas Douglas, fifth earl of Selkirk, possessed neither the powers of an HBC governor nor even those of deputy governor. Nor was he basically interested in the administrative minutiae of fur trading. What obsessed this passionately philanthropic man was the desire to re-settle families of Scots farmers in British North America. So he toiled at Beaver Hall in London every day from ten till six o'clock to master the details of a business he as good as owned. If, by improving the operation of the trading organization and meeting the opposition on equal terms he could further the welfare of his Red River settlement, no labor or expense was too great for him. And using the considerable experience, advice, and organizing abilities of Colin Robertson, a disgruntled, one-time Nor'Wester, Selkirk set out to force recognition from the North West Company of HBC fur-trading and settlement privileges alike in the *pays d'en haut*. One of Selkirk's and Robertson's warning shots to the opposition was the dispatch of a fur-trading party into Athabasca under the command of John Clarke, another former Nor'Wester and, as such, typically courageous, ambitious, and not a little vain-glorious.

The region into which Clarke led his traders and voyageurs was famed for an abundance of wild life that sustained Indian and European alike. The Peace River valley stretches several hundred miles eastward from the Rocky Mountains. Unlike that other great Mackenzie tributary, the Athabasca River, which flows through heavily timbered country, the Peace rolls through great prairies and is bordered by vast, park-like hillsides. The shape of the Peace River country has been likened to that of a huge goose rising out of the far regions of the province of Alberta.

> . . . the most northerly arable land in Canada, it hovers over the rest of Alberta. Its wing span extends three hundred miles, from Hudson Hope on the west to [Lesser] Slave Lake on the east. The great head and neck stretch out 150 miles north to Fort Vermilion . . . . It is the last frontier of the white man's settling . . . . Lucky were the fur traders . . . whose lot it was to be stationed in the Peace River Country instead of at Chipewyan or down the Mackenzie. They have described it in glowing terms. The region around Grande Prairie they had called the Buffalo Plains, but it soon became known as La Grande Prairie . . . . Dotted with groves of poplar, and covered with herds of buffalo and elk grazing everywhere on the luxuriant grass and vetches, what a grand prairie it was . . . For miles and miles this prairie extended . . . a glorious country of hills and streams and jewelled lakes, mostly prairie alternating with some heavily wooded sections. Off to the southwest, visible from elevation, rose the Rocky Mountains, a long jagged ridge made intensely blue by the distance. Over it all hung the scent of pea blossoms inviting the myriads of bees already gravid with honey . . . . There were all the other prairies of the great Peace River

country – Spirit River, Pouce Coupe, and High Prairie, and the famous stretch on the north side of the river from Dunvegan to [the town of] Peace River.*

Here grazed hundreds of small herds of buffalo and family groups of elk. Bears roamed the region in their thousands; indeed, the Peace River valley may have been the most thickly populated bear country in all of North America. In and out of the many clumps of brush next to the river wandered the lordly grizzly and its smaller, plumper cousin, the black bear, each feasting enormously on acre after acre of raspberry bushes and raiding the nests of wild bees. In marshes ringed by willows and thick with water lilies browsed the huge moose. And in the forests hemming in these prairies to the north lived the woodland caribou, a close relative of the Arctic caribou and distant cousin to the reindeer. All this big game meant plentiful supplies of meat: frozen, or dried, pounded, and then mixed with the huge, blue-black Saskatoon berries that grew over much of the valley; the end result, pemmican, sustained many a family through winters that commonly saw the temperature plunge to sixty below.

Despite the fact that he led veteran voyageurs and traders recruited in Montreal, John Clarke did not win the Athabasca campaign of 1815-1816. But then he found himself up against not only a tough but an enraged opposition. To the proud Highlanders who ran the North West Company – in particular those who served in its Athabasca Department – any invasion by members of an English company was intolerable. The basis of their wrath is easily explained: their rivals could perhaps lay legal claim to all lands and territories draining into Hudson's Bay, but it was Nor'Westers who had explored much of these lands and territories and then gone on even farther west and north into an entirely different watershed and there, at great individual risk and immense corporate cost, earned the right to trade. Now, at the whim of some *Lowland* nobleman – who

* MacGregor, J. G., *THE LAND OF TWELVE-FOOT DAVIS: A History of the Peace River Country,* Edmonton, The Institute of Applied Art, Ltd., 1952.

was in league with a *Sassenach* company – they were to be robbed of the fruits of their toil! At Selkirk's dictate, the Mackenzie watershed, which produced the finest furs to be found on the entire North American continent – and which was certainly no part of King Charles II's original grant to the Hudson's Bay Company – was to be taken over from its rightful owners!

So the bitter animosity of the numerous McLeods and McGillivrays of the Athabasca Department was directed against the hapless Clarke expedition. Not in any hurried, haphazard way. There was no need for that. Its Highland antagonists had trailed the expedition and noted its lack of winter supplies of pemmican. So the newcomers were allowed to establish posts on the Athabasca River, on Lake Athabasca, and on Great Slave Lake and garner a few furs. But, all the time, the Nor'Westers persuaded Indians to drive back game from the main waterways and ordered them, on no account, to supply meat or pemmican to any HBC man. Thus it was when Clarke, six of his officers, and forty-eight of his men canoed up the Peace River to establish posts that the full weight of their rivals' fury became apparent. Menaced, sometimes mauled by Nor'Westers, or made malleable by handouts of rum, no Indian would hunt for the men from the Bay or trade them food. So Clarke sent sixteen of the more exhausted members of his group back to Lake Athabasca, where they could at least subsist by fishing. They set off downstream in canoes, but the river froze, and they were forced to journey on foot. Individuals fell out by the way, one by one, to die of a mixture of cold, exposure, and hunger. Only three reached the safety of the HBC's Fort Wedderburn on the lake.

Meanwhile, Clarke kept forging farther and farther up the Peace River in a desperate search for provisions. Pressing on ahead of his brigade, he at last managed to buy food supplies from a band of Indians, although to do so he had to threaten to shoot Nor'Wester Alexander Roderick McLeod, who was trying to overawe these customers of his. But Clarke's effort was wasted. When he returned downriver to the brigade's camp, he found that three men had died of starvation and the remainder had, to save themselves, surrendered their trade goods to the Nor'Westers and signed a promise not to serve the HBC in Athabasca for a period of a year. Although he and a few companions held out for several weeks longer, living on whatever they could trap or shoot in the woods, Clarke, too, was obliged to give up the goods he had in return for enough food to take him back to Fort Wedderburn.

Undaunted by the defeat of his first attempt to establish the Hudson's Bay Company in Athabasca, Clarke returned the next year with fresh men and supplies. However, he decided to contest matters at Fort Wedderburn, a fort he built on an island directly opposite the Nor'Wester depot of Fort Chipewyan on the north-west shore of Lake Athabasca in order to draw away customers from the opposition. Here, thanks to excellent fishing grounds, Clarke and his men at least would not starve.

Clarke had little chance to win over customers. For one thing, his force of men was greatly outnumbered. For another, his opponent, a veteran wintering partner called Archibald Norman McLeod, waged a very successful brand of psychological warfare. Deliberately donning his scarlet military tunic and brandishing a long sword, this former major of militia in the War of 1812 regularly humiliated his competitors and showed everybody who was master of the situation at Lake Athabasca. He had a blockhouse built right next door to Fort Wedderburn so that his men could remonstrate with any Indians bringing furs to "the English." Abusing his authority as a Justice of the Peace under the Canada Jurisdiction Act of 1803, McLeod would, from time to time, force his way into the HBC post on some legal pretext or other, arrest personnel, and then free them on their guarantee to keep the peace for a six-month or a twelve-month period - all this in front of Indian trappers. McLeod's harassment of the HBC even included, on one occasion, impounding fishing gear and the men using it to obtain just about the only steady source of food available to them. Finally, A. N. McLeod, J.P. seized Fort Wedderburn itself. Clarke, forced to hand over trade goods to the value of £3,000, was sent as a prisoner to a Nor'Wester post on the Peace River. Thus ended the HBC campaign of 1816-1817.

The battle of wills between the two companies went on for four more years, disrupting the Athabasca fur trade, consuming huge sums of money, and costing a few men their lives. The whole thing developed into a sort of crazy contest of tit-for-tat: one side would capture and imprison a rival wintering partner, and then the other side quickly retaliated in exactly the same manner. The struggle for domination of the trade eventually spread beyond Athabasca. Each company began to ambush and waylay the other; there was even an occasion in 1819 when an HBC group mounted a three-pounder cannon on a barge, towed it north the whole length of Lake Winnipeg to the mouth of the Saskatchewan River, and there gave seven, east-bound Nor'Wester wintering partners the option of surrendering or being blown to smithereens. (They chose to live to fight another day: exactly a year later at the same location, one of these partners, plus sixty men, tried the same trick on HBC personnel, but failed.) And, like his rivals, Lord Selkirk had warrants issued for the arrest of this, that, and the next Nor'Wester. In the end, the tangled affairs and conflicting law suits of each company – which included the killing or dispersal of colonists at the Red River settlement, the destruction of the settlement on two occasions, and the capture by Selkirk, personally, of the great Nor'Wester depot of Fort William on Lake Superior – took years to sort out in the law courts of Upper and Lower Canada.

Nonetheless, by 1820, it was the Nor'Westers who were on the defensive. Slowly but surely, the HBC was establishing firm footholds in the watershed of the Mackenzie River; by patience, firmness, and a spirit of a fair dealing for which the Company was to become famous throughout the Canadian north, it was winning over customers. By the fall of 1820, the Company had not only enjoyed two good trading years on the Peace River but had even managed to establish a post almost as far west as modern Fort St. John, British Columbia. More importantly, the Company had greater resources of money than the Nor'Westers, whose finances were becoming greatly hampered by debt charges. In fact, a few wintering partners, dismayed and finally infuriated by the hard-line attitude taken by their Montreal agents in the struggle with

Selkirk, had already secretly contacted the Company in London, England, with suggestions of a merger. They knew only too well that the money that kept the North West Company going was the investment of *their* shares in the organization. Yet the Montreal agents insisted on maintaining a ruinously expensive trade war in their Athabasca Department, the balance sheets of which showed losses for several years in a row.

In 1820, yet another Highland Scot arrived in the Athabasca country. An HBC employee, he was destined to temper and then heal the feud between the rival fur companies. A Nor'Wester noted in his diary the arrival at Fort Wedderburn of "a stranger ... reputedly a gentlemanly man, but not formidable as an Indian trader." Never was a greater mistake made in reading character. The newcomer would contest and conquer such typical Nor'Wester "bullies" as Samuel Black, cozen and control proud characters of the likes of Archibald Norman McLeod, and go on from there to consolidate about a quarter of the North American continent into a commercial empire of unsurpassed efficiency. A distant relative of the great Alexander Mackenzie, he made a reality of his kinsman's dream of a unified fur trade whose operations ranged from the Atlantic to the Pacific and Arctic oceans. In particular, this man laid the foundations of a trading empire in the Mackenzie basin.

* * *

You didn't have to be a Scot to qualify for employment in the fur trade – but it helped. Generation after generation of Scottish fathers and mothers, uncles and aunts, grandfathers and grandmothers encouraged children to rise early, study late, save their pennies ("Many a mickle mak's a muckle!"), and worship a God who, youngsters were told, highly approved of the Protestant work ethic. Doubtless, these were the virtues taught the blue-eyed, red-haired George Simpson by the family of his grandfather, the Reverend Thomas Simpson, in the sea-side town of Dingwall in northern Scotland. Luckily for the lad, the illegitimate offspring of a ne'er-do-well son of the manse and an unknown girl or woman, he was given much more than these dour,

Calvinistic precepts; he was greatly loved and much mothered by his Aunt Mary, who refused the marriage offers of a persistent suitor until about the time George went off to London to earn a living. Whatever education in Latin and Greek wee George received, he was well-drilled in writing, grammar, and what in those days was called "commercial arithmetic." (A forty-year accumulation in HBC archives of Simpson's copy-book handwriting and economy-dominated thinking are fair proof that he was an attentive student.) Which is probably why one of his uncles, Geddes Mackenzie Simpson, offered him a job as a counting-house clerk in the firm of Messrs. Graham, Simpson, and Wedderburn, a partnership of sugar brokers located in Tower Street, London.

Now, the "Wedderburn" in the firm's name was Andrew Wedderburn, whose sister had married Lord Selkirk. (The HBC fort on Potato Island, a mile away from the Nor'Wester depot of Fort Chipewyan, had been named in honor of Wedderburn). Andrew may have been influenced by his brother-in-law, but being as canny a man as ever came out of the Scottish borders, he began to buy the dividend-poor but cheap-selling stock of the Hudson's Bay Company and within two years had become a member of the board of directors. As time went on, Wedderburn - who saw fit to rename himself Andrew Colvile - became the dominant member on the HBC's board of governors. His influence on its conduct of affairs was to overshadow Selkirk's, and, in fact, Colvile outlasted him by several decades. It was the shrewd, sharp-eyed Colvile who proposed that George Simpson be sent to North America. It was Colvile who argued that "to take care of our affairs" someone had to be available as a back-up man to the overseas governor, William Williams, who might be "dragged away" at any time by the Nor'Westers, thus depriving the HBC of a resident administrator and leader in its struggle against the "wolves of the North." It was Colvile who decided that whatever practical knowledge of the fur trade and of North America his protégée lacked was more than made up for by his qualities of energy and efficiency, courage and coolness.

Thus began George Simpson's long, enormously successful career with the Hudson's Bay Company. He himself thought in 1820 that he was going out to North America to spend either a summer or a year and then return to the business of buying and selling sugar. Instead, he was to become the Company's man, body and soul, until his death forty years later.

Simpson arrived in the *pays d'en haut* to find that William Williams had managed to elude opposition attempts to capture him, and the new man lost no time in following orders to organize the Company's 1820-1821 Athabasca campaign. In order to do this, he had two basic tasks to accomplish: supervising the preparation and despatch of a multiplicity of trade goods; and trading these items for furs in the teeth of a still fierce opposition. The detailed, strictly clerical nature of the first task partly explains why Simpson was Colvile's nominee for the job.

> As officer in charge of the Athabasca Department, Simpson had under his direction five districts: Peace River, Athabasca Lake, Great Slave Lake, Mackenzie River, and, beyond the Rocky Mountains, New Caledonia. For the trading posts, existing and planned, in this broad area, a truly fantastic variety of goods was required. These included augers and anvils, bayonets and beads and blankets, compasses and caps and chisels, dishes, files, grindstones and guns, hats, hinges and hooks, and so on through the alphabet to waistcoats and wires and worsted. Even the provisions for the Company servants in the Athabasca Lake district alone would stock a large grocery store. Among other things, they comprised thirteen 90-pound bags of flour, thirty pound of raisins, eighteen each of black and of green tea, seventy-six pounds of butter, and lesser quantities of such items as coffee, chocolate, mustard, nutmeg, pimento, pepper and sugar, as well as many gallons of wine, brandy, and rum. All of these, trade goods and provisions, had to be moved through hundreds of miles of wilderness, propelled by the strength of human muscle, supplemented only occasionally by a favorable wind. Every ninety-pound "piece" into which the goods were packed, had to be loaded and unloaded at each of a hundred portages, by strong arms, and carried the length of every portage on human backs. The control of such an undertaking was the responsibility of this young Scots business man. Yet keeping track of all the trade goods, of the three dozen brass thimbles destined to Fort St. Mary's [on the Peace River], of the

> cross-cut saw for Fort Wedderburn, of the small tin funnel for Berens House [on the Athabasca River], was perhaps the least difficult part of Simpson's job. Equally important was the employment and management of the men under his direction, the French-Canadian canoemen, the half-breed interpreters, the Indian guides, and finally . . . traders and clerks, who would be charged with converting the fruits of civilization into beaver and marten and fox skins. For when all these goods were moved to . . . the Athabasca Department, only then was it possible to begin the true business of the Hudson's Bay Company – trading for furs.*

Compared with such complex problems in logistics, which took him a few years to master and reorganize along much more economical lines, dealing with Nor'Wester bullies was almost something of a relief. Simpson adopted a courteous but no-nonsense attitude from the start. According to Simpson, the first opponent to feel the lash of his tongue was Simon McGillivray Junior, whose father was the Nor'Westers' corporate leader, William McGillivray.† He and several companions, each carrying a dirk and a brace of large pistols, turned up at Fort Wedderburn one morning. They had come to stop the construction of a stockade Simpson had ordered built to screen off from Nor'Westers in their nearby blockhouse trade proceedings between Indians and his men. Informed that the opposition objected to this move, Simpson walked out to the trench, carefully stood on the HBC side of it, and introduced himself.

"My name is Simpson," he said. "I presume yours is McGillivray?"

"It is," was the reply.

"I intend erecting these stockades from the corner of the bastion in a direct line to that stump," pointing to the remains of a tree that was understood to be the boundary line on the island between the two companies. "Pray sir, what are your objections?"

* Chalmers, John W., *Fur Trade Governor: George Simpson, 1820-1860*, Edmonton, The Institute of Applied Art, Ltd., 1960.

† Rich, E. E., ed., *Journal of Occurrences in the Athabasca Department by George Simpson, 1820 and 1821, and Report*, Toronto and London, The Champlain Society/The Hudson's Bay Record Society, 1938.

"I understand that they are intended to be run beyond the boundary line, which I shall not permit," was McGillivray's answer.

"We have no intention to encroach on what is understood to be the line of demarcation, nor shall we tamely submit to any encroachment on our rights," rejoined Simpson. "We are inclined to be quiet, orderly neighbors if permitted to be so, but are determined to maintain our privileges with firmness and shall promptly resent any injury or insult that may be offered."

The disbelieving McGillivray muttered sullenly, "Time will show."

So Simpson decided to rub the lesson in right there and then. At that moment, Boxer, his terrier dog, was on the other side of the trench playing with a stick right at the feet of one of McGillivray's companions, who seemed just about to give Boxer a swift kick in the rear. Simpson smiled and called out, "Come here, Boxer! You do not seem to be aware that you are committing a trespass."

McGillivray, thoroughly nettled by this, remarked, "We have no intention to molest your dog, Sir."

"Nor shall you his master with impunity!" shot back Simpson. Two hours later the stockade was in place. Word quickly spread throughout the Lake Athabasca area that the new man at Wedderburn was there to do business and wasn't going to be browbeaten by anybody.

Thereafter, the Nor'Westers tried various stratagems to oust the HBC from Fort Wedderburn, but Simpson countered these by arming his men and demonstrating every intention of meeting force with equal force. For a man who was never violent by nature or act, Simpson spent a lot of time that winter of 1820-1821 carrying a double-barrelled gun in his hand and a loaded pistol in his belt.

Despite the war of nerves, Simpson's main preoccupation was running the business of the Company in Athabasca, which involved a considerable amount of daily correspondence with all the other posts. Discipline was lax, if not non-existent, so he wrote his subordinates that servants who were disobedient, neglectful, dishonest or impertinent would be punished by fines, or imprisonment or, worst of all, reduced rations. Time and again he sent a typically polite but pointed communication to the effect that the trading system would soon be "improved"; instead of extravagance, which "seems hitherto to have been the motto," it would now be "oeconomy". Thus, "A sufficient quantity of fish should be taken in the fall for the maintenance of the people during the winter; they cannot expect meat except by the way of treat on particular occasions." "A quantity of thread will be sent you for the purpose of making nets for McKenzie's River." As for customers, "Every encouragement should be held out to them to renew their habits of industry; they should not be overloaded with Debt; short and frequent supplies [of trade goods] may answer better than giving them their full equipments in the Fall . . . ." There were instructions to be careful to lay aside a sufficient quantity of ammunition, rum, and tobacco for the summer trade. There were requests to "have the goodness" to prepare a full report of the district in question by the time he got there. There were reminders that beaver was the sole object of the HBC's mission and admonitions that it was by the number of packs of furs that the "Honourable Hudson's Bay Company can appreciate the Talents of their Traders."

No detail escaped Simpson's inquisitorial eye and ear. He found out that Duncan Finlayson, the trader in charge of a Peace River post, had a first-rate hunter who was also a highly temperamental character. "You should therefore study to please him, as in the event of a scarcity of Provisions, your existence in a great measure depends on him; I have made his little boy a present, which I think will please the Father." Simpson himself did not hesitate to curry favor with the wife of one of his own guides, on the grounds that she was the best interpreter at Fort Wedderburn. (As Simpson astutely advised his subordinates, "flattery is a very cheap commodity.") When it was hinted that Joseph Greill, the manager of Berens House on the Athabasca River, was addicted to the bottle, Simpson wrote to him, "this report I cannot believe until it is substantiated on conclusive evidence . . . a Drunkard you are aware is an object of contempt even in the eyes of the Savage race with whom we have to deal in this country." Even such a minor matter as sled dogs claimed his attention. Shocked to the depths of his thrifty Scots soul by their high price, mainly the result of their Indian owners' habit of eating those who had survived the winter in reasonable shape, he decided to issue an order that the Company should keep only bitches; under such an arrangement, "we shall always have a good stock of dogs, and can supply the people at fair prices." Always conscious of the terrible burden of expenses, Simpson even toyed with the idea of persuading Indians to capture caribou fawns and, as in Europe, breaking them in as draft animals for such tasks as hauling loads of fish or cords of wood, and dragging timber to sites where new posts would be constructed.

In short, he was the apostle of economy. To be wasteful of anything, more particularly to indulge in what Simpson called "luxuries" and other men thought of as ordinary European household supplies, grated like a rasp on his frugal sensibilities. "One would think," was his outraged reply to a post manager's indent for mustard, "from the quantity you order that it is intended to be used in the Indian trade." He was a typical, nineteenth-century captain of industry - with perhaps a major excuse for his penny-pinching ways. The scene of his endeavors was a raw, fierce wilderness, which called for an administrator as hard and as indomitable as the land itself. And, come Hell or high water, George Simpson was determined it would pay a profit. (With which sentiment, his kinsman, Sir Alexander Mackenzie, would have solemnly, but joyously, agreed.)

When that first busy winter was over, Simpson came out of Athabasca to find that the bitter war for the fur trade was over. At Grand Rapids on the Saskatchewan River, the very place where the rival concerns had in turn ambushed each other, a Nor'Wester clerk informed him that a coalition of the companies had taken place - and had to produce letters to prove it. Simpson's reaction was the small man at his pugnacious best: "I must confess my own disappointment that instead of a junction our Opponents have not been beaten out of the Field, which with one or two years of good management I am certain might have been effected."

His disappointment was short-lived. For Simpson, the coalition of 1821 was the stepping-stone to promotion and power. That same year, he was put in charge of all HBC operations north and west of the Great Lakes. (His Scots relatives must have rejoiced exceedingly at the news he was now earning £1,000 a year.) Five years later, he was Governor in Chief of every Company operation in British North America, which then included much of the modern American territories of Montana, Idaho, Oregon, and Washington. With power came prominence. In due course, he was knighted by a grateful Queen Victoria and lionized by British society. And when, in 1860, the then Prince of Wales (later King Edward VII) visited Montreal in the course of a highly successful North American tour, his guide, companion, and host was none other than the man jealous trade colleagues referred to as "The Little Emperor."

Of the many problems posed by the union of 1821, the greatest one was psychological. How could the proud, stubborn, mainly Highland-born "wintering partners" who more or less owned the North West Company be persuaded to work side-by-side as colleagues with the better-disciplined, law-abiding, largely English-born men employed as sales executives by the Hudson's Bay Company? For that matter, how were the men from the Bay going to feel about receiving into their midst those "lords of the lakes and forests" who had mocked, insulted, and physically injured many an HBC "field officer" and his servants in the course of almost fifty years of unremitting rivalry?

Nothing illustrates this difficulty more clearly than an eyewitness account by an HBC clerk of a banquet in 1821 at York Factory (a depot on Hudson Bay) attended by the principal men of each side, now created either Chief Factors or Chief Traders – regional and district sales managers, as it were.

> This first social meeting of the superior officers of the coalesced concerns – 73 men were present – in the great mess hall of the factory . . . 300 feet in length with its two long narrow tables, had some peculiar features, owing to the bitter feelings of the guests who had for many years been keen trade competitors, and sometimes personal antagonists in willing combat. The "proud Northwest bucks" – mostly Highland men – had been stalking

about the old fort, as haughtily as had been their wont at their own former headquarters for the interior, namely, Fort William, Lake Superior,* not trying to converse with the Hudson's Bayites. It was "dollars to doughnuts" - as the saying is - whether the entertainment would be a "feed" or a "fight." . . .

The two sections of the guests, at summons of the bell, entered the great hall in silence, and kept wholly apart until the new governor [George Simpson], moving in the throng with bows, smiles and introduction, brought about some conversation of hand-shaking between individuals, and ended by pointing at, politely, where he invited the guests to sit. It was hardly possible, in the circumstances, and owing to the number of guests, to avoid mistakes in this matter of seating, and in fact several unfortunate mistakes occurred.

Watching the banquet from a corner of the great hall, the scene was like some of those described in the "Legend of Montrose," a book I afterwards read. Men found themselves vis à vis, across the narrow table, who had lately slashed each other with swords, and bore marks of the combat. I noticed one Highlander so placed whose nostrils seemed to expand as he glared at his mortal foe, and who snorted, squirmed and spat, not on the table, but between his legs - he and his enemy opposite being as restless as if each were sitting on a hillock of ants. Their hate was real, yet as a spectator assisting in the ceremonies, I could not but feel a little tickle of the ludicrous. Another couple of good haters - a mobile-featured, black-eyed man of sinister aspect (under a suspicion of poisoning), and a pompous fellow, with neckerchief and collar, up to his ears - had lately fought a pistol duel across a camp-fire after night fall. Another was expected to take wine with his jailor opposite, who a few months before had imprisoned him, as a captive Northwester, in a dark cellar, where he had to inhale the premonitory fumes of brimstone. . . .

The situation was saved by the demonstrative - if not very sincere - comradeship of the several superior officers of the two sections, whose example others followed, though some continued to glare with fierce eyes at their former personal and official enemies. I feel bound to add, comparing small things

* Due to the establishment of a Canada-U.S. boundary line that resulted in Grand Portage's being located in American territory, the Nor'Westers had been forced to build another central depot on the shore of Lake Superior. They named it Fort William in honor of their corporate leader, William McGillivray.

with great, that the good effects of the fine wine used lavishly on this particular occasion, cannot be denied. Its action in helping to overcome rigorous discontent, reminded me of the effect of the spreading warmth of the summer season of this region in mitigating the winter harshness.*

It was just such a super-sensitive, stiff-necked group of men that Simpson took charge of, gently bullied, and flatteringly manipulated to build a bigger, better Hudson's Bay Company. He accomplished this by an open, friendly manner, an even-handed, impersonal administration, and a strict discipline. He was their boss, and in various ways he never let them forget this. But perhaps his greatest strength in their eyes was that he proved to be a money-making boss. Dishonest or inefficient traders and clerks were fired. Posts that did not pay their way were closed down, and those not well positioned were relocated. Personnel surplus to Simpson's idea of requirements were, if possible, shifted from trading activities to maintenance and supply work at the Red River Settlement; in four years he reduced the total of employees from 1,983 to 827. The wages of some employees were reduced by almost 50%. Perquisites and gratuities of various kinds were cancelled, and transportation systems and methods overhauled. Thus, when Chief Factors and Chief Traders noted that their HBC shares were paying increasing dividends, they responded more and more readily to the stream of directives that never seemed to stop flowing from Simpson's pen or mouth. Four years after he took over the reins of management, dividends were back up from 4% to 10% – and also paid a bonus of a further 10%.

Some Nor'Westers, of course, had to be given extra special treatment. Thus, the man Simpson called a "black-hearted bully" – to whom, incidentally, former associates had given a commemorative ring inscribed "To the most worthy of the Nor'Westers" – was quietly smuggled back into service again. Samuel Black was given the energy-absorbing tasks of, first, exploring and charting the headwaters of the Peace

* "Career of a Scotch Boy, who became the Honourable John Tod", edited by Gilbert Malcolm Sproat and printed in the Victoria *Daily Times* in issues from September 30th to December 23rd, 1905, reprinted in the *British Columbia Historical Quarterly*, Volume XVIII, Victoria, B.C., July-October, 1954.

River, and then developing trade with several particularly aggressive Indian groups along the lower reaches of the Columbia River. And Black's high-spirited colleague and friend, Peter Skene Ogden, another persecutor in times past of HBC personnel, was also admitted to employment on condition that he, too, went far west and there spearheaded HBC operations against American trappers in the wild mountain country through which the Snake River winds a devious way.

Governor Simpson bent every last member of the North West Company to his will, which he always thought of as an extension of the will of *the* Governor, and his Committee, in London, England.

* * *

While it may have been fate that sent a tiny gamecock of a man to take charge of Fort Wedderburn during the winter of 1820-21, it was no chance or accident that prompted the Governor and Committee of the Hudson's Bay Company to despatch someone to take charge of the Athabasca country. The drainage basin of the Mackenzie River had long been the source of the thickest, silkiest pelts obtained anywhere in North America. Had not the great Alexander Mackenzie written of the headwaters of the Peace River that "In no part of the North-West did I see so much beaver-work"? Why else would John Clarke, one of Simpson's predecessors in Athabasca, be sent in 1800 by his one-time Nor'Wester employers halfway down the Mackenzie River to establish the fur post of Fort Norman?

To the west of the Rocky Mountains, the historic Columbia Department of the former North West Company was made to yield money for many years after the union of 1821. In the mountain-and-plateau country of northern British Columbia, in the wooded parklands immediately north of the prairies, in the rocky, lake-and-forest reaches of northern Saskatchewan, Manitoba, Ontario and Quebec, and in Labrador, the HBC long continued to garner fur pelts. However, it was in the arctic and subarctic territories of its Mackenzie River District that a golden era of profits began for the Hudson's Bay Company.

Chapter 4

A Trading Empire

The drainage basin of the Mackenzie River is roughly triangular in shape, the apex touching the shore of the Arctic Ocean. Almost dead centre in the triangle is a huge inland sea, Great Slave Lake, from whose western end the Mackenzie "proper" rolls swiftly northward 1,100 miles to empty itself into the cold, grey waters of the Beaufort Sea. About another 1,100 miles south of Great Slave Lake is the base of this triangle, a base that extends from the Rocky Mountains all the way into northeastern Saskatchewan. So enormous is the Mackenzie basin that it also drains a great deal of water from British Columbia and from the Yukon Territory. Even if viewed from several hundred miles out in space, this watershed would be easily picked out. Between the brown mountain masses that wrinkle the face of northern British Columbia and the greenish forest and muskeg to the west of Hudson Bay flows the clear, black line of the Mackenzie's mainstream. To the southwest, thinner black lines indicate its principal tributaries – the Liard, Peace, and Athabasca rivers. Much more easily identified are the shapes of those other great sources of Mackenzie water: the flying saucer that is Lake Athabasca; the bent-nosed submarine that is Great Slave Lake; and, farthest north of all, the peculiar, near-swastika shape that is Great Bear Lake.

To most Canadians, all these thousands of square miles of territory are part and parcel of the "North," which starts somewhere just beyond the last section of cottage country and goes on forever. North is a land of snow-laden forests, frozen landscapes, and ice-bound seas. However, this notion of northern Canada needs some major qualification because, in fact, there are several "Norths."

Perhaps the most obvious North of all is the Arctic, commonly defined as that portion of the globe north of the Arctic Circle (66°33′ North latitude). But this description is nothing more than a sectioning off of part of the world on the basis of hours of sunlight. At the Arctic Circle, on one day the sun never sets and on another, some months later, it never rises; the farther north you go, the greater the number of sunny or sunless days. Some physicists delineate an arctic region by the frequency of an aurora borealis or some other evidence of behavior by the earth's magnetic field. Oceanographers have their own definition of "Arctic" based on the temperature and salt content of water. Then there's the North of the permafrost, the line beyond which some or all of the subsoil is permanently frozen; but the southern edge of the permafrost line wanders all over the place, and there is argument among scientists as to where this North ends – or, if you like, begins. Even temperature can be a deceptive guide. Cold is the condition most widely associated with the term "Arctic": yet, while the Arctic is cold *longer* than regions much farther south, the areas of extreme cold in the Northern hemisphere are located south of the Arctic Circle.

Advances in scientific knowledge are part of the trouble in defining "North" or "Arctic" accurately and precisely. Where these words once had one or more simple, historic-political definitions, they are now more a group of concepts and attributes that are still being refined. Nonetheless, temperature does provide an extremely useful and widely accepted definition that dispenses with the vagueness of "North" and uses the still somewhat imprecise but handier divisions of "Arctic" and "subarctic." As a general rule, trees will grow where the average July temperature is above 50° Fahrenheit (10° Celsius for purists). Whatever the severity of winter cold, with a summer respite of such proportions, trees will grow, even flourish. Thus, primarily as a result of climate, there is a limit to their growth: beyond this line are the treeless hills, rolling tundra, and mountainous islands of what can be called "the Arctic."

This "tree line" is by no means regular in outline. From the Mackenzie's delta next the Beaufort Sea it wanders across the Northwest Territories and northern Manitoba to the southern shores of Hudson Bay and then continues across Quebec and on into Labrador. All locations along the line have similar temperatures (bearable in summer if you don't mind mosquitoes and black flies, bloody freezing in winter). Thus, summertime on the Mackenzie – 300 miles of which lies north of the Arctic Circle – is as uncomfortably hot as that in northern Manitoba or Quebec. Perhaps more interestingly, as fur traders noted, the tree line was a cultural boundary beyond which an Indian rarely ventured. Beyond it lay what Samuel Hearne called "the barren lands," a graveyard of glacial debris and the haunt of a traditional Indian enemy, the Eskimo.

If the lichen-and-moss wilderness of the barren lands west of Hudson Bay awed the few HBC traders who tried to penetrate westward, the landscape of the subarctic region northwest of the prairies baffled the Pedlars from Montreal. Here are heavily-timbered stands of spruce, pine, fir, and larch thickly sown with streams, swamps, and lakes. In an area ten miles square, there may be a hundred lakes distinctive enough to be mapped and another hundred smaller ones that all look much alike. Rivers only added to an explorer's bewilderment because they often open out into huge bodies of water and just as frequently narrow again to rocky channels with as many twists in them as there are threads on a screw. Shores and banks are sometimes sandy, sometimes naked outcrops of Pre-Cambrian granite. Away from the water are woodland clumps of poplar and birch or low-lying land speckled with shallow, reedy ponds, plus patch after patch of spruce-ringed muskeg stretching off into infinity. And the principal inhabitants of this region are mosquitoes and blackflies that swarm in clouds that grow denser and deadlier the farther north you travel.

There are, too, the frustrations of climate. Here, summer days can be suffocatingly hot, the nights a weird kind of twilight. And the farther north you go, the longer winter becomes – almost nine months by the time the delta of the Mackenzie is reached. The Mackenzie trader's world was primarily a sea of snow, driven by polar winds into swells, waves, and enormous breakers. Swirling around and against a fur post, these finally cascaded over the highest palisade, and the buildings within often disappeared under its white flood. Frosty mists hide horizons in smoky clouds for weeks on end, and over all hangs a grey, fog-like darkness that serves as daylight. On clear nights, the Northern Lights flash and flit through the sky like demented devils at play, adding a vaguely disturbing dimension to an already hazardous existence. In his post journal, one old-time HBC man summed up his experience of the vagaries of the sub-arctic winter in verse penned by candlelight on a lonely December 31st.

> If New Year's Eve the wind blow south,
> It betokeneth warmth and growth;
> If west, much milk, and fish in the sea;
> If north, much storm and cold will be;
> If east, the trees will bear much fruit;
> If northeast, flee it man and beast.*

A more reassuring discovery traders made in what most of them called the Northwest was the gentle, almost timorous reception they got from its human inhabitants, the "babiche people." They lived by babiche, ingenious rawhide skins, cords, and traps made from the hides of deer and caribou. Masters of travel by snowshoe and sled, expert hunters and fishers, they knew how to survive in the treacherous sub-arctics and eagerly shared this knowledge with the new-comers. These Chipewyans, Slaves, Beavers, Copperknives, Hares, and Dogribs could be violent and murderous towards each other, but rarely to a European. And after a century or more of ferocious oppression by Cree middlemen armed with HBC muskets and busy exploiting a monopoly situation, these Athapaskan-speaking people were only too glad to deal with men who sold them goods at much fairer prices.

Furs, of course, were available by the million. Beaver was found wherever there was a combination of water and the animal's favorite foods, the bark of aspen and poplar, although a beaver will cheerfully munch on the bark of just about any available tree. What especially stimulated beaver

* Cameron, Agnes Deans, *The New North: Being Some Account of a Woman's Journey Through Canada to the Arctic*, New York City, D. Appleton and Company, 1909.

hunting was that the animal's pelt could be used for fur, the tail eaten as a delicacy, and castoreum, a thick, oily, secretion in a pair of castors or glands, sold to the perfume trade as a fixative. In addition, if the flesh was not eaten, it could be fed to sled dogs. Muskrat, whose pelt had long ranked alongside seal and mink, was plentiful in marsh and swamp. Members of the bloodthirsty weasel family – mink, martin, and fisher – abounded in rocky, forested uplands, and another, particularly murderous member of this family the ermine, haunted the northernmost woods. Hares and rabbits and, more importantly to the trader, their great enemy, the fox, were also present in great numbers. As time passed and the great male mania for fur hats was superseded by a female frenzy for fur coats, neck pieces, hat trimmings and edging for elaborate gowns, other furs were supplied by the lands bordering the Mackenzie and its tributaries. The great staples were lowly skunk, which was marketed under the classier name of "black sable," a Russian term for the smallest member of the marten family; lynx, the king of all cat furs; wolf; black bear; and even hare and rabbit, dyed and disguised under such labels as seal, ermine, fox, and lynx.

It was this particular treasury of natural riches that the Hudson's Bay Company acquired by the union of 1821.

* * *

After the fierce fights of the days of competition, relative peace and contentment descended upon the Company's Mackenzie Department. Under Simpson and successive governors, the HBC's attitude was strictly businesslike, but beneficent. There was no grabbing for every last pelt. In fact, in 1825, Simpson and his senior colleagues probably introduced the principle of conservation to North America when they forbade the killing of beaver in summer months. Persuading customers to give up the habit – fostered by the Nor'Westers – of exploiting an area's fur resources right down to the thinnest or mangiest pelt was quite a job. But, in time, most Indians resumed their ancestors' ancient practice of maintaining a steady number of fur-bearers in any one district. For his part, Simpson circulated a memorandum to the effect that ". . . as the posts must, to a certain extent, be maintained to preserve the Indians, who could not now exist without ammunition and other necessaries, expenses cannot be curtailed in proportion to the returns." As a result of this and other farsighted dealings, relations between Company and customer settled into a pattern where each came to trust and help the other. At least, this was true of business relationships between most traders and trappers. Simpson, however, had little or no regard for Indians. A few of them, he admitted, were "manly" and "honorable." The rest were "cowardly" or "treacherous." All of them, he decided, "must be ruled with a rod of Iron to bring and keep them in a proper state of subordination . . ." They just weren't businesslike enough for his taste. Firmness, tempered by a certain fairness, was all he ever personally offered his clientele.

Not the least unsuccessful of Simpson's efforts to improve profits was his adoption of an old French and Nor'Wester practice. He issued an official invitation to his senior personnel to take unto themselves a "country wife," to marry *à la façon du pays*. As he delicately phrased it,

> Connubial alliances are the best security we have of the goodwill of the Natives. I have therefore recommended the Gentlemen [the Company's commissioned officers, that is, Chief Factors and Chief Traders] to form connections with the principal Families immediately on their arrival, which is no difficult matter as the offer of their Wives & Daughters is the first token of their Friendship & hospitality.*

As in times past, this marriage arrangement worked admirably. Companionship or love apart, an Indian woman with a European husband enjoyed immense local prestige (and, incidentally, a great deal of ease and comfort since she was spared the laborious, "feast-or-famine" type of existence usually endured by her fellow tribeswomen). Her presence in a post ensured that relations and friends - usually amounting to an entire Indian band - would bring in furs and meat.

Both from a Company and from a personal point of view,

* Rich, E. E., ed., *Journal of Occurrences in the Athabasca Department by George Simpson, 1820 and 1821, and Report*, Toronto and London, The Champlain Society/The Hudson's Bay Record Society, 1938.

Simpson himself was a fervent believer in socializing with local belles. Long before his own connubial alliance with a cousin, he had enjoyed, as he quite indelicately put it, "several bits of brown." Simpson was not only cynical about his frequent use of Métis women as handy objects of sexual gratification but could be brutally two-faced about the institution of marriage with a non-European. In 1831, HBC man Colin Robertson visited the Red River Settlement with his Métis wife, to whom he was devoted all of their long life together. Simpson, by this time married, rebuked Robertson for the impropriety of bringing his "bit of Brown with him to the Settlement this Spring in hopes that she would pick up a few English manners before visiting the civilised world . . . I told him distinctly the thing was impossible, which mortified him exceedingly."*

Whether at a post on the Peace, on the Athabasca, the Liard, or on the Mackenzie itself, the routine of trading hardly varied. A dog-sled team would arrive at some rocky point or headland, near which was clustered a group of HBC buildings. Warned by the barking of the post's dogs, the manager, together with his clerk and other servants, would be sure to greet his visitors and invite them to have a warming "mug-up" of strong tea. (By 1828, the Company had gradually eliminated the use of liquor as a bribe or even as a preliminary to trading.) In the "Indian Hall" or trading store, there followed a courteous exchange of local news – births, marriages, deaths, illness, a little gossip, perhaps some scandal – and then rather rambling conversations about the weather, the behavior of various animal species, and the vagaries of hunting. Conversation dealt with anything but the reason for the visit to the post. Finally, some little thing triggered the trading session. Perhaps it was a mention of Carcajou the wolverine, the truly vicious member of the weasel family who loves to rob traplines of captive animals and bait alike and delights in breaking into trappers' cabins, wrecking their contents and fouling every-

---

* Rich, E. E., ed., *Colin Robertson's Correspondence Book, September 1817 to September 1822,* Toronto and London, The Champlain Society/The Hudson's Bay Record Society, 1939.

thing with urine. Whatever the cause, onto the counter were laid various bundles of peltries.

If it was the early winter period, the bundles contained plenty of fisher, otter, and mink, for these fish-eaters are easily taken before midwinter temperatures solidify every stream and small river. There would be many muskrat, because their underground dens, sealed off from the waters of slough and pond by successive frosts, are natural traps. Next to muskrat, the commonest skins – not particularly welcomed by the trader – were rabbit and hare, which at least assisted some fur manufacturer to produce clever imitations of grey fox, chinchilla, and seal. The trapper may have included a few, particularly fine, wolf pelts, but usually the trader only paid attention to those of an Arctic species with brown or white coloring. What the trader did want to see was the really cold-weather furs: a creamy white ermine, soft as a swan's feather; a silver-white Arctic fox; or, best of all, the silver-grey lustre of a beaver, whose hard-wearing pelt almost equalled that of a buffalo hide. It was this combination of beauty and durability that made beaver so desirable and, as a result, the basic unit of trade. To trader and customer alike, a beaver skin was as good as money in the bank.

In order to barter to the satisfaction of each party, some easily-understandable system of exchange was necessary. Historically, this varied from wampum (shell beads) to discs made of ivory or bone, to porcupine quills, to wooden sticks, to metal tokens of brass, copper, and, in recent times, aluminum. But, for several centuries, the basis of all of this "currency" was a beaver pelt known as "One Made Beaver," meaning a prime winter pelt taken in good condition. A customer bringing his furs to a Company post received, in exchange, a number of Made-Beaver tokens approximately equivalent to the value of furs presented to the trader. With his tokens, a customer reduced or eliminated any debts he had in the Company's books for credits advanced at some time or other, and then bought whatever items he and his wife needed. From somewhere within his caribou-skin jacket, he pulled out a bit of paper with some characters on it as a reminder to purchase various items. First and foremost would be tobacco and ammunition, and then perhaps metal traps. And so some tokens would pass across the counter. After some dickering and good-natured argument – normally instigated by the customer's wife, who enjoyed driving a bargain and expected the trader's clerk to have a sliding scale of prices – there would be further purchases.

By the nineteenth century, there was an incredible selection of goods to choose from. Every HBC post was a mini-department store. While it stocked the customary blankets and beads, pots and kettles, guns and knives, its shelves were also filled with dozens of other items: bonnets, caps, and coats; lengths of cotton, flannel, muslin, and silk; shawls, shirts, and shoes; buttons, needles, and threads; fancy vests, and at least a dozen types of trousers from summer tweed to tartan wool. Hardware ranged from adzes and files to frying pans, saws, and vises. If you owned a horse, you could buy bridles and brushes. If you were a gardener, there were seed packets of carrot, onion, lettuce, turnip, and radish. (Vegetables grow to huge proportions in the brief but hot northern summer with its long hours of daylight.) If a cook, you could get cinnamon, cloves, nutmeg, olive oil, pepper, and pimento. Parents could buy slate pencils and ebony rulers for their children and indulge them with chocolate, peppermint lozenges, barley sugar, and raisins. There was even a drug section that offered Turlington's Balsam – which claimed to cure most ailments from aching ankles to varicose veins – as well as liniment, castor oil, plasters, Epsom salts, and so forth. And for those with a taste for gracious living, there was Java coffee and a choice of Congo, Hyson, or Suchong tea.

Bartering items manufactured in British factories or harvested in Asia was a fairly simple proposition. What was far from simple was transporting several thousand tons of trade goods into the Mackenzie District year after year and bringing out less weighty but equally bulky bales of furs. This problem in logistics is an old Canadian story – the effort to traverse enormous distances and overcome some of the worst terrain in the world. In the case of the HBC's Mackenzie District, it was a story as epic in its way as creating the Canadian Pacific Railway.

In school, every Canadian is subjected to descriptions of the contruction, use, and history of the birchbark canoe, together with overly romanticized descriptions of its colorful French-Canadian paddlers. (They *were* the workhorses of the trade; but, as many diarists confirm, they were also verminous, foul-mouthed, carnally-minded and, in winter, most adept in managing to do little or nothing in the way of work.) But how many Canadians would recognize a York boat if they saw one? Historians and other writers have given the Nor'Westers' freight canoe, the *canot de maître*, its smaller, swifter *pays-d'en-haut* version, the *canot du nord*, and their swarthy-skinned, song-singing voyageurs a pride of place and a glamor that has totally overshadowed that of their much less picturesque but more efficient rival.

There is a close family resemblance between the York boat and the fishing craft that ply out of the tiny harbors of Scotland's Orkney Islands. The likeness is not accidental. In the eighteenth century, something like 75% of the HBC's employees were Orkneymen. So they shaped their new watercraft with its high bow and stern after that of the old, itself a lineal descendant of the Viking longboat. With a figurehead added, shields lashed to the side and a colored square sail rigged to the mast, a York boat could easily pass for a smaller version of a Norse raiding ship. Painstakingly built of carefully selected spruce, these "inland boats," as they were often termed, were designed to draw as little water as a canoe. And this they did superbly. As early as 1803, Alexander Henry (the Younger), a Nor'Wester wintering partner, noted in his diary meeting an HBC brigade of boats "neatly built and painted and sharp at both ends; they are propelled by 4 oarsmen and a steersman and carry 45 packages . . . averaging 80 lbs each." This must have been an early version of the York boat because not very many years later its crew was six to eight rowers transporting seventy pièces of 90-lbs weight. Of course, portaging a York boat was hard work – slashing a wide trail through the bush, clearing away rocks, and then cutting up trees to provide log rollers on which to manhandle the boat past a particular water obstacle. Other than that, the York boat outperformed the birchbark canoe on several counts. A far sturdier craft, it forged ahead during lake storms that kept canoe brigades weatherbound for days on some beach, was unaffected by the razor-sharp action of ice floes, and handled better under sail. Above all, the York boat was more economical, outclassing even the massive *canot de maître*, because it carried more cargo in proportion to crew.

Thanks to George Simpson's sharp eye for cutting expenses, after the union of 1821, the York boat displaced the canoe. With his zeal for economy, Simpson improved matters further by ordering the construction of even-larger versions of this flat-floored craft drawing only 2 feet of water that increased individual cargo loads to 3 and 3 1/2 tons. And what cargoes these were. Bales, boxes, kegs, packing crates, and tea chests were commonplace items of cargo. Occasional passengers were carried, along with their trunks, cases, hat boxes, surveying equipment, bedrolls, and other camping paraphernelia. (In artist Paul Kane's case, his paints, brushes, canvases, and easels, too.) Some cargo manifests defy belief: catherdral bells for the Church of St. Boniface in the Red River settlement; wheeled carriages and pianos (disassembled, of course); six- and nine-pounder artillery pieces; livestock, including cattle; and on one occasion a couple of young, bad-tempered buffalo being exported via York Factory. Plus an item that spelled the ultimate doom of the York boat: cast-iron boilers for river steamers.

Two very important articles carried were a medicine chest and a tool box. Equipped according to a formula worked out by the Company's medical advisors, the former contained.

> 2 pounds of Epsom salts
> 1 dozen purgative powders
> 1 dozen vomit powders (in little packets)
> 1 small bottle of smelling salts (for passengers with delicate sensibilities)
> 1 or 2 rolls of sticking plaster
> 1 lancet
> 1 pair forceps.*

---

* *Trader King, as told to Mary Weekes: The thrilling story of forty years' service in the North-West Territories, related by one of the last of the old time Wintering Partners of the Hudson's Bay Company*, Regina and Toronto, School Aids and Text Book Publishing Company Limited, 1949.

There was a book of instructions with these drugs, but none of the men, not even the guide (the most experienced navigator), could read English. However, everyone knew the dosages by heart, and the guide was the doctor. If a man needed medicine, he had to fetch a panniker of water, into which the guide poured what he thought to be the right dosage. In order to discourage shamming, the patient had to drink his medicine right in front of his medical advisor. If a tooth had to be extracted, the patient was placed against the side of the York boat, the "doctor" got a firm grip with the forceps, braced one foot against the bulwark – and hauled away. Boils were lanced with a swift jab; a dab of rum and a patch of sticking plaster were the only concessions to germicide. As an HBC officer observed of medical practice while in transit, "Everything was managed very well," adding casually, "And a jack-knife was all that was required for an amputation."* But it was the tool box that was the real life-or-death item. To be stranded in the wilderness for want of tools and spare parts was to invite hardship, possibly death. So every York boat carried a hammer, axe, handsaw, assorted nails and chisels, a brace-and-bit, a gimlet, caulking irons, oakum, a pail of tar, a couple of yards of spare canvas, needles and thread, a half dozen awls, and babiche (rawhide) for sail trimming. Again with safety very much in mind, room had to be found for other basic equipment: bailing pans, extra sail and rigging, spare oars, and the removable steering rudder.

Sturdy and economic though the York boat was, it wasn't the long-term answer to the twin problems of doing business in the far north: immense distances, and a boreal climate. High-quality furs were obtainable all over the Mackenzie basin yet, at best there were only four to five months of the year in which northern waterways were ice-free. And, in the latter half of the nineteenth century, a third, complicating factor began to make itself felt: the more numerous the Company's clientele became, the heavier the total freight load each way. In the last half of the nineteenth century, by which time HBC personnel had established trade outlets all the way down the Mackenzie and also westward into Yukon

* *Ibid.*

watersheds, the Company had customers and employees working three to four thousand miles away from its major supply depot of York Factory on Hudson Bay. By 1876, Athabasca alone, the most southerly trade district in the Mackenzie Department, was consuming some 35 tons of trade goods annually.

For several decades, part of the answer to the Company's northern transportation headache was the use of huge, clumsy scows, which were worked up and down the lower reaches of the Athabasca River by crews of sweating, swearing boatmen. Indeed, this particular section of Mackenzie headwaters became the funnel through which percolated the entire trade between northern Canada and much of the rest of the nation.

Freight was taken up and down the Saskatchewan River to Fort Edmonton by York boat or, after the CPR reached Calgary in 1882, shipped by rail between Winnipeg and Calgary and then transported over the 200 miles between that centre and the Fort by plodding teams of ox-drawn carts and covered wagons. (In time, a rail line reached Edmonton.) Between Fort Edmonton and the half-dozen warehouses at Athabasca Landing on the Athabasca River lay a hundred-mile, rough-and-ready trail through forest and swamp that converted boys into men overnight. Cauldron-like mud holes were the least of the travelers' difficulties. The weather blew alternately hot and cold. Much of the time, there were violent rains, which brought forth armies of mosquitoes and flies. Freighters could protect themselves to some extent with netting over their head and neck, but nothing of the sort could be done for the draft animals. They were tortured, sometimes to the point of madness, by the incessant stabbings of the mosquito and the infinitely worse bite of the horsefly, which literally tore out a chunk of flesh and left bloody trickles oozing down an animal's shoulders and flanks.

At the tiny, bustling centre of Athabasca Landing, the gateway to the North, workers in several shipyards framed, planked, nailed, and caulked local timber to produce dozen after dozen of that outsize packing crate of a watercraft called an Athabasca scow. It was no thing of beauty. Flat-bottomed, square at bow and stern, about 50 feet long and 8-12 feet wide as it flared towards the top, it resembled nothing so much as a monstrous, unlidded coffin. But a scow could carry five tons of merchandise and be broken up later for construction lumber downriver. If, by some mischance, it didn't survive the various rapids of the lower Athabasca, then, after rescuing its cargo, all that had been lost was some easily-replaceable timber.

And so, for a time, there came into existence yet another type of voyageur, the scowmen of the Athabasca River, who labored for twenty, thirty, or forty dollars a month, plus meals, plus an endless supply of moccasins. Mainly Cree in origin, although a few were Métis, these tall, dark-haired, dark-skinned men spent their winters hunting and trapping, ranging the woods with a sled loaded with not much more than blanket, bait, and bacon, and snowshoeing up to 40 miles a day from trap to trap. In summer, they risked life and limb as scowmen. For the lower section of the 250-mile run from the Landing downstream to Fort McMurray at the junction of the Athabasca and Clearwater rivers was, at best, tricky, at worst, treacherous. As it rolls steadily northward, the river cuts an ever-deepening trench out of the clay of the northern plains until the water finally meets bedrock. There are accumulations of boulders, causing first riffles, then rapids. Finally, rock layers form the bed, and the river cascades over the edges of broken strata. Sometimes narrow intricate channels lead through or past these obstacles. At other locations, a scow had to be steered right over the drop. If the water level was high, you usually got past in one swooping rush, and then the crew stick-handled like mad with their pole-like oars to recover steerage way. If it was a low-water year, the men "worried" their craft over, which meant poling it free of the ledge, the scow awkwardly tilted half in, half out of the water, and the whole operation attended by jarring shudders as the scow gradually scraped free of the ledge. As an old Athabasca boatman remarked on being told that the HBC had framed a time-table covering every movement of all north-bound traffic, "Yes, yes; the Company it makes laws, but the river he boss."

Getting through the most fearsome broken water on the

river – naturally known as Grand Rapids – was a mixture of calculated chance and sheer hard work. Here, the river tumbles over a natural dam of rocks and goes on to drop 30 feet within a quarter of a mile. While some of this could be navigated by very careful manoeuvring and much desperate work with the poles, unloaded scows had to go down a shallow side channel in order to avoid the massive boulders that jutted out of the frenzied waters in the middle of Grand Rapids. So cargoes had to be unloaded at the head of an island in this stretch, trundled down its long length in push cars that ran on wooden rails encased in sheet metal, and then reloaded. The remaining miles to Fort McMurray kept everyone on the alert because ten rapids had to be run. Two of these were so vicious that it was normal practice to run a scow brigade through them one vessel at a time, each worked by a double crew of the most skilful voyageurs in that brigade. Tourists bold enough to journey north via the Athabasca noted that two natural phenomena produced much profanity from scowmen – in English, for the Cree language lacks swear words: mosquitoes, and the Boiler and Long Rapids.

At Fort McMurray, most of the scowmen were paid off. After idling about for a day or two, most crews in a brigade set out on the long walk back to Athabasca Landing, up hill and down dale, around muskeg or right through it, and across mile after mile of fallen timber. Still, this was a great deal easier than "tracking" fur-laden scows back upriver, a chore some of the men had to perform.

In certain circumstances, tracking a canoe can require a fair degree of effort; hauling a York boat against the current was always a tough proposition; but dragging a clumsy lump of a thing like a scow sapped a man's strength and stamina. Harnessed like horses and roped to hawsers on board, scowmen pulled these deadweights every last one of the 250 miles back to the Landing. It wasn't the action of hauling itself that was difficult but finding a footing in wet sand, on slippery clay banks, and amidst pebbles, stones, and rocks. That, and keeping ropes clear of brush entanglements or clumps of willow. Or stopping every now and then to wade into the river and rock the damn scow off a sand bar. Or being thrown on your face when a rope snapped, retracing your steps downstream to the place where the steersmen had stopped the scow drifting, and starting the whole bloody business all over again. Those who walked back to Athabasca Landing did so in a leisurely week. Those who, season after season, labored from first to last light at the end of a track line, with only five-minute rest spells every hour, took anywhere from three to four weeks to complete the trip. They didn't do it just for the extra wages or the four, hearty meals a day. The prime motivation was esprit de corps: the proud knowledge that *they* were the real Athabasca scowmen, the élite of the northern waterways. More simply, and echoing an earlier breed of voyageur, they could say, with ample justification, "*Je suis un homme du nord*."

Downstream at Fort McMurray, awaiting the arrival of scow brigades, was the other part of the answer to HBC transportation problems: the steamboat.

In 1882, John Walters of Edmonton signed a contract with the North-West Navigation Company Limited - a Hudson's Bay Company subsidiary - to build three scows at Athabasca Landing to take supplies, equipment, and machinery downriver and on past Fort McMurray to Lake Athabasca itself. Here, at the HBC's great northern depot of Fort Chipewyan, the equipment and machinery would be utilized to assemble a steamboat 135-feet long and 24-feet wide with a cargo capacity of 200 tons. In order to do so, 25,000 lbs of parts, which included a 2½-ton boiler, had to be hauled the hundred or so miles over the crude track to the Landing and then freighted seven hundred miles downriver to Fort Chipewyan by scow.

At Edmonton, the offloading of the bits and pieces of the future sternwheeler *Grahame* had caused quite a buzz of interest. But this was nothing to the impatient, winter-long, curiosity of Fort Chipewyan's inhabitants as workmen, supervised by the vessel's first master, Captain John H. Smith, whipsawed, cut, trimmed, and shaped lumber to build what seemed to those accustomed to canoes, a replica of Noah's Ark. Rumor had it that the lower deck would be so large that, even when loaded with pièces, there would still be enough room left for the crew to dance a Red River jig. And, wonder of wonders, the shipbuilders were also creating a floating hotel; it would have not only dining tables, linen, cutlery, and real dishes, but individual bedrooms complete with bath! Finally, in the fall of 1883, construction was finished. A small mountain of cordwood was carried aboard, and Engineer Littlebury's stokers heaved logs into the furnace until the boiler's pressure gauges recorded a respectable head of steam. Captain Smith gave the order to cast off, and the S.S. *Grahame* thrashed away across the blue-and-green surface of Lake Athabasca to a snug winter berth on the south shore. The following year, the sternwheeler was voyaging back up the Athabasca River to Fort McMurray, where huge piles of goods awaited transport "down north" as far as a series of rapids on the Slave River – the last obstacle to navigation between Lake Athabasca and the Arctic Ocean – or all the way up the Peace River to the white-capped barrier of the Vermilion Chutes.

But not even a sixteen-mile series of rapids was going to stop HBC men from using steam to defeat time and distance on Mackenzie waters. In the summer of 1885, Captain Smith, the architect and master of the *Grahame*, left Athabasca Landing in charge of three scow-loads of materials with which to construct another steamship, this time for use on the Mackenzie "proper." As bad luck would have it, the scow bringing the boiler of the future S.S. *Wrigley* broke up in one of those vicious rapids towards the end of the Athabasca Landing-Fort McMurray run (thus immortalizing this broken water with the name "Boiler Rapid"). In fact, the river was so high that year that Smith and his men made the rest of the trip with half loads, coming laboriously back upstream for the remainder. After a long delay, the boiler was salvaged and, under Smith's watchful eye, transported to Fort Smith*, and ultimately installed in the new, propeller-driven craft. Exactly one year later, the *Wrigley* enjoyed the distinction of being the first steam vessel to cross the Arctic Circle anywhere in the world and had the added

* Fort Smith, located at the northern end of the rapids, was named by Chief Factor Roderick MacFarlane – yet another relative of Alexander Mackenzie – in honor of Donald Smith, Commissioner of the HBC 1870 to 1874, perhaps better known to Canadians as one of the group of men who organized the construction of the CPR.

distinction of being the first such ship to reach the ice-fringed mouth of the mighty Mackenzie.

A few years later at Fort Smith, the Company succeeded in launching yet another steam vessel, appropriately named *The Mackenzie River*. Its construction was dogged by misfortune. Twice, landslides came close to dumping the partially-finished vessel into the Slave River; then a fire burned down the blacksmith's shop and, with it, the doors, windows, and interior wood finishings laboriously brought in from the "outside." Indeed, when she finally sailed, with her went the carpenters, still busy sawing and hammering. But *The Mackenzie River* turned out to be the queen of all the river boats. Almost half again as large as the *Wrigley*, she had a hull reinforced by steel to ward off the blows of ice and floating timber and was further safeguarded by five water-tight compartments. Propulsion was effected by a pair of engines that provided an average speed of 10 mph. There was stateroom accommodation for twenty-two passengers, and cargo capacity enough to carry goods and provisions to keep the northernmost posts going for a full twelve months.

* * *

The first tourists to travel on the S.S. *Mackenzie River* were a vacationing American schoolmarm and a friend. Something about this trading empire that struck Miss Cameron was the John Bunyan flavor of several of its placenames.

> The names given by the old fur-traders to their posts make the traveller think that in these North lands he, a second Christian, is essaying a new Pilgrim's Progress. At the south entry to the Lake [Great Slave] we are at Resolution; when we cross it we arrive at Providence; away off at the eastern extremity is Reliance; Confidence takes us to Great Bear Lake; and Good Hope stretches far ahead down the lower reaches of the Mackenzie.*

This wasn't at all that caught the teacher's attention. At the depot of Fort Simpson on the Mackenzie itself, Miss Cameron came upon further clues to the character of old-time HBC men. In a library at this depot were leather-bound volumes

* *The New North: Being Some Account of a Woman's Journey Through Canada to the Arctic*, New York City, D. Appleton and Company, 1909.

of a decidedly uplifting nature: copies of Virgil's works, and all of Voltaire and Corneille in original, unabridged editions. A set of Shakespeare showed signs of "hard reading." For the very serious-minded, there was a copy of Robert Burton's *The Anatomy of Melancholy*, plus a huge tome entitled the *Annual Register of History, Politics, and Literature of the Year 1764*. Nearby on the same shelf was a little book with a big title: *Death-Bed Triumphs of Eminent Christians, Exemplifying the Power of Religion in a Dying Hour*. Next to that was a collected edition of *The Spectator*, a magazine whose literary tastes and moral strictures had heavily influenced the growth of professional journalism and periodical writing in Great Britain. There must, however, have been a few lighthearted characters in the Company's employ who ordered reading material from home, because Fort Simpson's library did house one or two items not devoted to a strictly literary education or the merits of self-improvement. Next to Burton's learned treatise on life and literature was a much shorter, more genial treatment in the form of *The Philosophy of Living or the Way to Enjoy Life and Its Comforts*. And two small books tucked in among the heftier tomes were *Actors' Budget of Wit and Merriment*, and *Memoirs of Miss A—n, Who Was Educated For a Nun, with Many Interesting Particulars*. Miss Cameron admits that she and her companion were sorely tempted to steal this wicked-sounding item but decided not to give way to sin, adding sadly, "And ever since we have regretted our Presbyterian upbringing."

Miss Cameron also records meeting typical HBC old-timers. One had come to the depot of Fort Chipewyan on Lake Athabasca in 1863. Actually, she first encountered him while reading the Fort's journals, in which there were references to "Wyllie at the forge," "Wyllie straightening the fowling-pieces," "Wyllie making nails," or "This day Wyllie made a coffin for an Indian." In person, he turned out to be a kindly, quiet-spoken Orkneyman. Although born in the Scottish isles, he'd never seen a city in either the Old or the New Worlds. Shipping out to Hudson Bay in a Company sailing vessel, he'd journeyed all the way to Fort Chipewyan via the inland waterways of the Hayes, Saskatchewan, and Athabasca rivers. Torontonians, wrote Miss Cameron, think the "hub of their universe is their capital on Lake Ontario."

Well, one day, a young man from Toronto somehow turned up at Fort Chipewyan and happened to ask the old blacksmith, "Came from the Old Country, didn't you? What did you think of Toronto?" He was mortified to be told, "Naething at all! Ah didna see the place!"

In all the forty-five years he'd labored at the Fort Chipewyan forge, the limits of Wyllie's world had been Fort McMurray on the Athabasca to the south, Fort Smith on the Slave to the north, Fond du Lac on Lake Athabasca to the east, and the Vermilion Chutes of the Peace River to the west. And he was clearly unconcerned with modern "progress." The fact that he'd never seen a building taller than two storeys, a railway, or a telephone never gave him a moment's unease or regret. Wyllie was quite content with his place in the scheme of things. He had spent a happy lifetime perfecting himself in the arts of gunsmithing and blacksmithing. It was a source of great satisfaction to him that his hammers and augers, planes and chisels were fashioned out of pig iron that arrived in the New World as ballast in the holds of sailing ships beating their way into Hudson Bay. And Wyllie was immensely proud of the fact that much of the ironwork of the S.S. *Grahame* had been wrought on *his* forge. Another such Company craftsman-cum-technician was William Johnson. Without ever having seen a light bulb switched on, he knew all he needed to know about electricity from reading and personal experiment. Proof of this was an electric-generating plant he first drafted and then constructed at Fort Simpson on the Mackenzie "proper." Johnson then decided to teach himself all about clocks and ultimately became known as "Father Time" throughout the Mackenzie Department, regulating and repairing every timepiece in the region. On one occasion, he solved the problem of a defunct watch by extracting parts from three old ones and re-assembling the various bits and pieces within the original case to make the faulty timepiece work.

Journals and diaries in Hudson's Bay Company archives are full of such stories. For instance, at the time of union in 1821, the HBC's stalwarts were a mixture of former Nor'Wester wintering partners and HBC "field officers" who became, in the terminology of the times, "Commissioned Gentlemen" (Chief Factors or Chief Traders). Some of these men were given authority as absolute and territories as tremendous as those of Roman proconsuls, yet remained surprisingly uncorrupted by power. John McLoughlin, who superintended all trading activities west of the Rocky Mountains, positively dismayed HBC management by the kindly, helpful manner in which he treated American pioneer families who "invaded" those portions of his Pacific empire that ultimately became the states of Oregon and Washington. The example of the "Father of Oregon" was followed by his successor, the principal architect of a Canadian province. A McLoughlin subordinate for many years, James Douglas, too, learned to administer a commercial-cum-political empire on the Pacific coast. Ultimately, after being obliged to serve both the Company and the Crown at one and the same time as the governor of the HBC-operated colony of Vancouver Island, the march of events in the Pacific northwest forced Douglas to give his loyalty to one or other. After a forty-year career based almost as much on maintaining law and order as on pursuing profits, he decided to become the first governor of the new, this time British-owned colony of British Columbia. As such, he displayed a concern for official affairs that sometimes baffled and irritated his one-time employers, the HBC governor and committee in London. East of the Rockies, at Fort Edmonton, was another ex-Nor'Wester. Using nothing more than sheer force of personality, John Rowand literally ruled the prairies with a firmness that commanded respect even from the ferociously proud Blackfoot nation and their equally haughty enemies, the Assiniboine and Plains Cree.

The Company's Mackenzie District had its share of such characters. Like many a colleague serving elsewhere in Canada, they were usually the sons of humble families living in the glens and isles of northern Scotland. They were equipped for life with little more than a sound elementary education in reading and writing, a smattering of classical knowledge, and a profound knowledge of the virtues itemized in the Shorter Catechism. (One father is reputed to have advised a son setting off for England and other foreign parts, "Dinna forget to keep up yer Presbyterian worship! It wullna stop ye from sinnin', but it wull prevent ye from enjoyin' the expeerience!") One of the finest descriptions of the average recruit was written by an English-born trader, who observed of the Company that

> Although itself an English corporation, its officers in the fur trade country are nearly all natives of Scotland and the Orkneys. More than one consideration, probably, contributed its weight in the selection of this nationality as its working representatives, viz., their proverbial shrewdness and propensity for barter; their generally vigorous physique and love of adventurous life; a steady perseverance in the attainment of an end; close economy, and the giving and receiving of the last half-penny in trade; and, above all, a certain Presbyterian honesty begotten of the Established Kirk.
>
> Successful applicants for place in the Company's Service . . . are enlisted invariably at an early age – generally from sixteen to eighteen – having first passed a rigid scrutiny as regards educational attainments, moral character, and, above all, physical build . . . The nominal term of enlistment is five years, although the more direct understanding is that the applicant shall devote his life to the trade – an event which happens in nearly every instance, the style of living being calculated to unfit him for active duty in any other vocation. . . . he is generally sent to pass the first five or ten years of his apprenticeship in the extreme northern districts of Mackenzie River. . . . This is done that he may at once be cut off from anything having a tendency to distract him from his duties. . . .*

With a background of thrift and frugality, together with an inherent toughness of body and soul, these youngsters were superb material for HBC work in northern Canada. As apprentices, they were honest, hardworking, and ambitious. What they may have lacked in social polish and sheer joie de vivre – characteristic of the occasionally English or French-Canadian compatriot – was more than made up for by their diligence and fortitude. An outstanding example of the breed, a long-time Mackenzie River employee, was Robert Campbell.

---

* Robinson H. M., *The Great Fur Land, or, Sketches of Life in the Hudson's Bay Territory,* New York City, C. P. Putnam's Sons, 1879.

A tall, broad-shouldered, black-haired lad, born and bred on a Perthshire farm, Campbell came to Red River in 1830 to help manage the Company's new experimental farm. Like many an HBC servant before him, he shipped out of Stromness in the Orkney Islands aboard a Company vessel and made the long Atlantic trip to Hudson Bay. Even late in August, this vast, inland sea was obscured by fog and choked with ice, on which Campbell and several fellow apprentices "amused ourselves by playing football and shooting at targets . . . sometimes at seals."* A leisurely inland voyage of about 1,300 miles by York boat brought him to the Red River Colony at the junction of the Red and the Assiniboine rivers. After a year or so of supervising grain growing and sheep herding, Campbell developed a bad case of itchy feet and begged Governor Simpson, who was wintering at Fort Garry, to give him a job as a fur trader. Recognizing a sturdy soul when he saw one, Simpson appointed Campbell to the Mackenzie River Department. Characteristically, the Governor's advice to the young man was, "Now, Campbell, don't you get married, as we want you for active service!"

Thus began for the farmer-turned-fur trader seventeen years of subarctic life. Using posts on the Mackenzie River as bases, he regularly negotiated the appalling rapids and whirlpools of the Liard River to work his way into the western watershed of the Mackenzie and establish posts in territories that had never been visited by Europeans. Time and time again, his journeys brought him face to face with groups of natives who had never seen a European and whose first instinct was to kill the intruder. Campbell reports that he became quite accustomed to laying down his weapon and holding out empty hands in sign of peace – all the while facing drawn bows and arrows. On no less than three occasions, he probably owed his life to a particular Indian woman. The first of these occurred in 1838.

That summer, Campbell and a party of eight men in two canoes labored up the Liard River, which was much the same thing as ascending a hill of water two thousand feet high and seven hundred miles long. The names of some river

* Public Archives of Canada, M.G. 19, A. 25, "Journal of Robert Campbell."

obstacles – Hell's Gate and the Devil's Portage – give some idea of the force and fury of Liard waters. So does the fact that the boats, each manned by a double crew, took seven days to ascend the mainstream of the Laird and seven hours to descend the same distance.

For once, Campbell and his men ate well, meeting with "many moose and reindeer [caribou], grizzly, black, and brown bears, and beaver whose cutting of poplar trees along the river looked like clearings made by axemen." Eventually they reached a long narrow sheet of water (Dease Lake) at the head of the river and immediately began building a fort. Pushing on with three companions to explore the country to the south, Campbell had the unusual experience of observing the different seasons of the year in the course of one day. "We passed over the shoulder of a lofty, snow-clad mountain. By 6 o'clock at night, owing to the difference in altitudes of the descent, we had experienced the 4 seasons – the snow of winter above, then the sprouting grass of spring, next the luxuriant vegetation of summer & finally at the foot of the mountains, the ripe strawberries of autumn." Two days later, he arrived at a fur-trade convention being held by numerous mountain and Pacific-coast tribes alongside the Stikine River, whose harvest of spawning salmon kept the hundreds of Indians present well-fed for weeks on end. While many of these natives traded with Russians at the mouth of the Stikine, Campbell met a few Indians who had voyaged far to the south in their great sea canoes and had met John McLoughlin and James Douglas of the HBC's Columbia Department. And it was at this huge gathering that Campbell first met

> The Chieftainess of the Nahanies [*sic*] . . . over which she & her father, a very old man, held sway . . . She commanded the respect not only of her own people, but of the tribes they had intercourse with. She was a fine looking woman rather above the middle height and about 35 years old . . . She had a pleasing face lit up with fine intelligent eyes, which when she was excited flashed like fire . . . At our first meeting, she was accompanied by some of her tribe & her husband, who was a nonentity . . . she soon gave us an evidence of her power. It appeared that . . . a

gun, firebag,* small kettle & axe had been taken from my party by the Indians [in the encampment], and as they were indispensable for our return journey to Dease's Lake, I was much annoyed. The Chieftainess saw there was something wrong & on discovering the cause, she gave some directions to 2 young Indians, who started off to the great camp, & who to my astonishment soon returned with the missing articles.

Campbell maintained an appearance of calm in a camp swarming with Tlingit middlemen who regarded his arrival as a potential threat to their profitable business of trading Russian goods for furs with the interior tribes. But his new friend, "the Chieftainess," was extremely uneasy – or had wind of trouble in the making – and hustled him and his men away from the great rendezvous and onto the trail back to Dease Lake. She even accompanied them for some miles, urging Campbell not to stop until nightfall in case some young Tlingit warriors, with the sanction of their chief, took it into their heads to do a little bloodletting. Campbell doesn't say so in his journal, but Russian traders were present at this rendezvous. He later reported verbally to one of his superiors that the senior among them "received him with apparent politeness and introduced him to three other ragged Officers like himself, and treated him to a Glass of Whiskey. It was, however, evident that they were jealous of his appearance in that Quarter." So his benefactress doubtless had very good reason for getting him out of the encampment with all possible speed.

That winter, Campbell and his men were gradually starving to death at their new fort. From the very start, Dease Lake had yielded remarkably few fish, barely enough to feed all hands at any time. As early as August, the daily yield began to diminish. Yet it was only large catches of fish that carried many a northern fur post through the winter. To make matters worse, there was an unpredictable shortage of game for miles around in every direction. Campbell records that "Our efforts all winter to procure a bare living were never relaxed. We were scattered in twos and threes trying

* The name for an ornamental bag containing a pipe and tobacco, steel and flint, or brimstone matches.

with nets and hooks for fish, & with traps, snares & guns for any living thing, bird or beast, that came in the way. Everything possible was used as food: trip de roche [a lichen that filled the stomach, although it barely appeased hunger], skins, parchment [window coverings], in fact, anything." Then, in February, came a brief respite. The Nahanni chieftainess paid Campbell a visit after discovering that one of his men, temporarily encamped at the north end of the lake, had died of hunger and exposure. A quick look at the fort's inhabitants told her all she needed to know. She ordered her servants to cook the best provisions being carried on her sleigh, and an hour or so later dried salmon and fresh caribou meat were being served up under her personal direction.

Later that same night, Campbell again witnessed the power this woman exercised over her followers. He and a companion were sitting in the fort's main room chatting at a table when yell after yell suddenly broke the evening quiet. Several Nahannis burst into the room, some loading their guns, while others seized HBC weapons from wall racks and then rooted around in search of powder and shot. Spluttering with rage, several of the intruders finally levelled their muskets at the two traders. Roused by the commotion, their leader rushed into the room and commanded silence. She questioned and queried and finally found out the instigator of the riot. Walking up to him and stamping a foot on the ground, she repeatedly spat in his face. Whatever the reason, real or imagined, for the attack, it was speedily disposed of. In Campbell's words, "Peace and quiet reigned as suddenly as the outbreak had burst forth. I have seen many far-famed warrior Chiefs with their bands in every kind of mood, but I never saw one who had such abolute authority or was as bold & ready to exercise it as that noble woman. She was truly a born leader, whose mandate none dared dispute."

Fighting off starvation became a more or less regular feature of Campbell's life for many of the winters he spent in the watershed of the Liard, a killer of a river that he sensed "had long been the dread of people in McKenzie's River [Department]." Even travelling in summer was no guarantee of food in the cooking pot. Two years after his Dease lake adventures, in the months of June and July, he trekked so far into the mountainous north that he climbed out of the Mackenzie watershed, entered another drainage basin, and there discovered a river he named the Pelly (a tributary of the mighty Yukon River). While the early part of the trip was, in his words, "luxurious," thanks to an abundance of beaver, trout, and game, "for three days on this trip we had neither the luck to kill nor the pleasure to eat . . ."

At this point in his career, Campbell had pioneered the trade northwest of the Mackenzie watershed and penetrated into Pacific-coast territories drained by the Yukon River. For years, he and his men had battled their way up and down the Liard, which Campbell continued to curse as a river "under a spell of some malediction, a source of endless mishaps and confusion." In the fourteen years he'd been on the river, it had claimed the lives of fourteen fellow employees. Then, at long last, fortune seemed to favour the manager of Fort Selkirk on the upper waters of the Pelly River. Governor Simpson, who had long kept a fatherly eye on Campbell and corresponded directly with him from time to time, sent him permission to explore the unknown lower reaches of the Pelly, which both Simpson and Campbell suspected was a tributary of the Yukon. (Their joint suspicion was duly confirmed when Campbell turned up at the HBC's Fort Yukon far down the main river.) The result was that Campbell found he could use a recently established overland route between Fort Yukon and the Mackenzie's mainstream. Using the Porcupine and Peel rivers to thread a way through the mountains that lay between the Yukon and Mackenzie watersheds, he was able to employ several hundred miles of the mainstream itself as a trouble-free highway to and from his supply depot of Fort Simpson at the junction of the Liard and the Mackenzie.

Shortly after his return to Fort Selkirk, Campbell encountered Indian rivals who put him out of business. A band of Chilkats, long angered by HBC interference with their well-established role as middlemen between local Indians and Russian traders, systematically wrecked the fort's contents and ran off its inhabitants. Indeed, it was only due to the protective efforts of a couple of friendly Chilkats, whom he'd used on occasion as letter carriers, that Campbell's life was spared. So late in the fall of 1852, he found himself not only a refugee back at Fort Simpson but forbidden by Chief Trader James Anderson, the head of the Mackenzie Department, to re-establish Fort Selkirk and remove, in Campbell's words "the stigma the Indians will cast on the Character and Bravery of the Company's officers." Quite unmollified by his own promotion to the rank of Chief Trader – which came through the day after handing in his report of the catastrophe – Campbell fastened on snowshoes and set off with a dog team for Montreal to argue matters out with the Governor, Sir George Simpson himself.

Between November 30th, 1852, when the Mackenzie ice was finally fit for travelling that year, and March 13th, 1853, by which time he had reached the Minnesota village of Crow Wing, Campbell actually set a record for snowshoe travel – approximately 3,000 miles. To Campbell, the whole thing was perfectly natural and amounted to little more than a longer-than-average hike. Stopovers along the way at various HBC establishments were pleasantly refreshing pauses. Christmas and New Year feasts were enjoyed at the depot of Fort Chipewyan on Lake Athabasca. At Fort Garry on the Red River, he dallied for fully five days "enjoying the luxuries of civilized life and society." Yet life on the trail wasn't all that bad, at least not according to Campbell and two voyageur companions. The exertion of snowshoeing made the entire body comfortably warm, and at every meal halt a huge fire of brushwood and branches kept the winter chill at bay. At night, a buffalo robe spread on spruce or balsam boughs made an excellent mattress; then, with a couple of blankets, plus another robe on top, even "when the thermometer is ranging at from 20 to 30 to 50 degrees below zero, men will sleep soundly, and perhaps to their ideas, more comfortably than in the best bedroom in the first hotel in London." The only problem seems to have been stretches of deep snow atop river ice, which caused the dogs to flounder and the sled to tip over.

Campbell journeyed through the United States by horse and buggy, stagecoach, steamer, and railroad, reaching

Montreal four months and a day after quitting Fort Simpson. (In the process, he had been on the waters of rivers draining to the Arctic, to Hudson Bay, to the Gulf of Mexico – via the Mississippi – and to the Atlantic. For that matter, when he had been ousted from the Yukon by the Chilkats, he had been in the continent's fifth watershed, the Pacific slope.) The transition from wilderness life to the lights, crowds, and industrial hustle and bustle of Montreal completely bewildered him. "I felt like a man entering a new world of which he knew nothing." So did his reception by Sir George, who wouldn't hear of his going all the way back to Fort Selkirk to "square up with the Chilkats." Campbell was reminded that twenty-three years of inland service without leave was reputation enough for any man and that, in any case, he was too well-known in the north for anyone to fancy him a coward. Further, remarked the Governor, had he not been requesting home leave for some years past in order to "fetch back a wife to cheer my long dreary winters in the North"? Well now, did he want to get married or didn't he? Were a few bad-tempered Indians more important than properly-cooked meals and a woman to keep the bed warm on cold nights? Thus, three days later, Campbell was off on his travels again, this time by rail to New York City, by steamer to Liverpool, and by train to London. According to his home newspaper, the *Perthshire Advertiser*, the journey from Fort Selkirk to the hub of empire was a trip of 9,687 miles.

While in the United Kingdom, Campbell concluded two very important items of personal business. First, he became engaged to his boyhood sweetheart, Miss Elleonora C. Stirling of Comrie, Perthshire. His second item of business was with the Arrowsmiths of London, a famous map-making concern. In 1852, this firm published a map of British North America. On it, much of the western watershed of the Mackenzie River appeared as a complete blank, although a dotted line, marked "Cook's River" and following roughly the course of the Yukon River, was strangely reminiscent of Peter Pond's inspired, but wishful, thinking as the main waterway of the Pacific Northwest. Thanks to Campbell's exploration and his London visit in 1853, Arrowsmith's 1854 edition of the same map was a much more accurate version. The headwaters that feed the mighty Liard were clearly outlined. The Peel and Porcupine rivers, which enter the Mackenzie just above its delta, were revealed as waterways originating deep within the Mackenzie and Ogilvie Mountains. And, at long last, the erroneous "Cook's River" was removed from the map of North America and replaced by, once and for all, the Yukon River.

There is much more to the life of Robert Campbell. He married his Elleonora – who, in due course, presented him with two sons and a daughter – and took charge of the Company's Athabasca District. In the year in which Confederation occurred, Campbell made the big step from Chief Trader to Chief Factor, which entitled him to a portion of HBC profits to the extent of 2/85ths in lieu of salary. And in later years, retired from the Company, he was a gentleman of leisure, whose many friends included the best-selling novelist, Rider Haggard. (Did Haggard perhaps get the idea for *She* from Campbell's account of his journey to the headwaters of the Stikine River and his rescue there by the Nahanni "chieftainess"?)

In his memoirs, Campbell glosses over his resignation from the Company in 1871, a resignation prompted by a drastic change in the HBC's financial dealings with its commissioned officers. It's odd that a man with his strong character and stern sense of duty should be so offhand, almost secretive, about a development that affected the future of every HBC officer in Canada. But then, by this date, Campbell was old, tired, subject to bouts of illness, and concerned for the daily well-being of a young family, whose members were utterly dependent upon his modest retirement income. Perhaps he felt it was not in their best interests if he engaged in argument and dissension with a head office whose aim and outlook were markedly different from those of the London governors and committees of George Simpson's day and age.

* * *

It has frequently been stated that the existence of the North West Company came to an abrupt end in 1821 when it was forced to merge with the Hudson's Bay Company. Yet this was not the opinion of those working in the fur trade of that time. In the blunt words of one old-timer, "The Hudson's Bay Company never amounted to a damn until the North West Company joined it!" In the more polite words of Edward Ellice, spoken to a British parliamentary investigation of the Hudson's Bay Company in 1857, "From these different arrangements [the union of 1821] sprung the present Hudson's Bay Company, which is more in fact a Canadian company than an English company in its origin."*

Edward "Bear" Ellice knew what he was talking about. This opinion was given by a man who, at one time, made his living by supplying trade goods to the North West Company, enjoyed the business confidence and personal friendship of Sir Alexander Mackenzie, and had a notable hand in arranging the union of the two rival companies in 1821. Several facts confirm the truth of Ellice's opinion. Strange as it seems, the Hudson's Bay Company was, quite literally, a reconstituted North West Company operating under a different name.

First and foremost, the act of Parliament of 1821 had reaffirmed the Hudson's Bay Company's ancient, chartered right to the trade of all lands whose waters drained into Hudson Bay. This legislation also granted a monopoly of all fur trading in any territory not granted to the Hudson's Bay Company and not belonging to the United States, to any foreign power, or to any British provinces in North America. In other words, about one third of the entire continent was made the private preserve of the Hudson's Bay Company. However, this latter privilege was *also* granted (for 21 years) to William and Simon McGillivray and Edward Ellice, directors of the North West Company.

Second, Nor'Wester wintering partners had always owned individual shares of their company, whereas HBC men had not (with the exception of a brief attempt by Lord Selkirk, on the advice of ex-Nor'Wester Colin Robertson, to apportion shares in HBC trade to senior personnel). Under the Act of Union, "clear gains and profits" of the united concern were to

* Public Record Office, London, *Report of the Select Committee of the House of Commons on the Hudson's Bay Company* . . ., 1857.

be divided on the basis of one hundred shares: forty became the property of the wintering partners of the former North West Company and certain senior personnel in the HBC.* These forty shares were subdivided into eighty-five shares, two for each Chief Factor, one for each Chief Trader, and seven shares were set aside for a retirement fund (doubtless based on the Nor'Wester practice of allowing a wintering partner to draw half-pay for some time after his retirement). This arrangement was guaranteed by the issue of a deed poll, a legal document that named the Chief Factors and Chief Traders, the men in the field who would own, jointly, 40% of the profits of the reorganized Hudson's Bay Company. They were the executives entirely responsible for the day-to-day operation of the fur trade. And this list of personnel is very revealing. Of the twenty-five Chief Factors, fifteen were ex-Nor'Westers; of the twenty-eight Chief Traders, seventeen were former Nor'Westers. Not only were thirty-two of the first fifty-three commissioned officers of the new HBC former competitors, but, in the next ten years, only two HBC men were promoted to the supreme status of Chief Factor. Two decades later, only three of the original HBC Chief Traders had been elevated to a chief factorship, while ex-Nor'Wester colleagues continued to rise to this rank. Small wonder Ellice testified before members of the British Parliament that "the present Hudson's Bay Company is more in fact a Canadian company than an English company in its origin"!

But Ellice made an even more significant remark before he left the witness stand. In 1857, there were many interests ranged in opposition to the Hudson's Bay Company and they dredged up stories of discreditable deeds and malodorous misdeeds committed by the Company since 1670. Ellice replied that he knew nothing of conditions a century or more earlier. Nor did he know anything of the various, specific charges levelled against the original owners of the Com-

* Of the remaining sixty shares, twenty were to go to William and Simon McGillivray and Edward Ellice, twenty to the Hudson's Bay Company, ten were to be invested in the united concern as an emergency fund, five further compensated the McGillivrays and Ellice for the loss of their business agency, the North West Company, and five were to be used by the HBC to fulfil an obligation to the heirs of Lord Selkirk.

pany, the "Gentlemen Adventurers of England trading into Hudson's Bay." When pressed for further comment, Ellice defeated his questioners and stunned the committee of enquiry by stating, "Nothing that previously took place in the affairs of the Hudson's Bay Company can at all have reference to what has been the conduct, or the management of the Company, for the last forty years."*

He was absolutely right. Of the original band of "Adventurers of England," nothing survived in this reorganized Hudson's Bay Company. No courtiers sat at its board meetings. No London directors wrote futile pleas that the trade be "pressed inland." Every vestige of the concern created by King Charles II in 1670 had been wiped out. The men, the methods, the energy, the devotion, the unconquerable will to succeed belonged to the Nor'Westers and to their lineal descendants, the Chief Factors and Chief Traders who owned a piece of the Company. The "lords of the lakes and forests," as the American, Washington Irving, once admiringly christened the Nor'Westers, had *not*, in that same writer's words, "passed away."

No, the North West Company did not die in 1821. It breathed its last in 1893, when the Hudson's Bay Company, slowly but surely converting itself into today's highly diversified, computerized corporation, cancelled the rights of its commissioned officers to a share of the profits. With this cancellation, with the regression of the officers to employee status, ended a way of life that was distinctively Canadian. With this cancellation disappeared a breed of outstanding executives, men who were motivated by loyalty to an organization that was largely what *they* made it. According to an HBC employee who joined the Company well before the coalition of 1821 but lived long enough to witness its renaissance in later years,

> At its best, probably there has not been any business organization so well devised as the Company to earn profit for the stockholders and partners. . . . The recruiting grounds of Quebec and Scotland, and, as time passed, the increasing number of men of mixed blood in the Indian country, helped the

* *Ibid.*

> fulfilment of this condition. Employment was scarce in these countries, and many of the people were adventurous. That accounts for the rank and file, but it is less easy to understand the uncommon proportion of ever-available, extraordinary men in the higher ranks of the service – a proportion not reached, as persons say who are better able than I am to judge of such a matter, in any modern industrial or commercial organization.*

The conversion of a fur-trading company into a modern corporation with several different sources of income is a story that fills a volume or two. But three early indications of this change stand out like landmarks. The first was the sale of the Hudson's Bay Company to the International Financial Society (IFS).

In 1863, with the reluctant knowledge and approval of the governor and committee of the Company, the 285 proprietors of the HBC were invited to sell their shares, each with a par value of £100, to the IFS for £300 at a time when the London Stock Exchange was quoting these shares at about £200 each. Most of the shareholders promptly sold their holdings. Just as promptly, the IFS increased the Company's capitalization of £500,000 (based on fur-trade earnings of about £300,000 a year) to £2,000,000 and announced a sale of HBC stock with a par value of £20. Prospective buyers were offered the opportunity to share in an anticipated exploration of the furs, fertile lands, forests, and minerals of the Northwest by the HBC and by the British and Canadian governments "in accordance with the industrial spirit of the age, and the rapid advancement which colonization has made in the countries adjacent to the Hudson Bay territories." All of this was rather optimistic thinking, but was fair, square, and above board for all that: the Company's properties and its ownership of at least one million square miles of territory covered the extra capitalization. What was wrong with the share issue, or rather its promotional propaganda, was that it raised expectations that could not be met, at least not in the western and northern Canada of the 1860s and 1870s. The immediate result of the sale was an

* "Career of a Scotch Boy, who became the Hon. John Tod," memoirs edited by Gilbert Malcolm Sproat and published in the Victoria *Daily Times*, Sept. 30-Dec. 23, 1905.

entirely new set of money-hungry shareholders (about 1,700 all told), whose frustration at low dividends, no stock splits, and none of the potential profits implied in IFS publicity made it extremely difficult for successive HBC boards of directors to operate the Company solely on a fur-trade basis. The ultimate result of the sale was that shareholder dissatisfaction, sometimes downright hostility, became a major factor in forcing the Company to make money by selling land and moving into the retail-store business.

The second landmark event was the sale of Rupert's Land and the rest of the Northwest to the Dominion of Canada on December 1, 1869. The negotiations leading to this were as long and involved as those that had led to the IFS purchase, and once again the Company was subjected to considerable pressure from the British government. In fact, this time around, terms of sale were more or less imposed on the HBC by the British government. When the shareholders first heard the details of the terms, they raised the roof. A miserable £300,000, the retention of all Company establishments, together with 1/20th of all land in a "fertile belt" bounded by the 49th parallel, the Rocky Mountains, the North Saskatchewan River, and the waters connecting between Lake Winnipeg and Lake of the Woods? This was all they were going to get for a business that netted a steady 4% annually in fur sales, operated a hundred or so depots and posts, maintained a shipping fleet – and had retained a third of the North American continent within the British Empire! It was almost an insult that the Canadian government would also buy several tons of telegraph wire stored at York Factory, which was as far as the International Finance Company had got in its early attempts to "open up" the West by means of a highway and a telegraph line.

In the end, the governor and committee of the HBC succeeded in ramming this bitter medicine down the shareholders' throats in the same determined way it had been rammed down theirs. It was clearly impossible to resist the combined wishes of the British and Canadian governments. But the real losers were the Company's commissioned officers. It's one thing for a businessman to have one of his investments go sour. It's quite another when a fur trader's private territory, which gave him his sole income via his one- or two-share holding, becomes public property. For one of the conditions of the sale of its former territories to Canada was the loss of the Company's fur-trading monopoly. Now, any individual pedlar or company could move in and trade for pelts. And long before the century came to a close, there was serious competition from several firms. In addition, the transfer of the entire Northwest to Ottawa's jurisdiction made it possible for the government of Sir John A. Macdonald to approve the construction of railways beyond the Great Lakes and encourage settlement by surveying and selling land. Settlement and fur trading had long been opposing pursuits, so there would undoubtedly be a steady decline in fur returns in future years.

But still worse news was in store for the Company's commissioned personnel. The third landmark event was the Deed Poll of 1871, which heralded the end of the old Nor'Wester practice of shared ownership of the fur trade.

The Company's executives in the field had not been consulted about the sale of the organization in 1863, nor were they advised of the sale of Rupert's Land and what Ottawa called the "North-Western Territory." The former angered them, the latter stunned them and left a legacy of distrust and great unease. Under the terms of the Deed Poll of 1821, the London Committee was under no legal obligation to consult its Canadian officers on such matters. However, many an officer felt that the new shareholders had at least a moral obligation to share with their overseas partners something of the £300,000 received for the surrender of territory in light of a future deterioration of their income and status. (The Chief Factors and Chief Traders were acutely aware that they had had to force the governor and committee to guarantee them a minimum £275 a share in the years 1865-1869.) For that matter, didn't these same partners have some claim to the £45,918 recently paid by the U.S. government in reparation for the loss of the Company's Oregon territory back in 1848? But the shareholders would not allow the Company to offer, let alone give, any part of these monies. Instead, in 1871, the best the governor and committee could do was to renew the partnership agreement begun in 1821 and try to sweeten things with pension-fund monies and also a portion of the U.S. dollars received from the sale of an HBC agricultural venture in Oregon.

Recognizing the intent of the 1821 Deed Poll that an officer's share should be worth about £500 a year – although few annual returns had averaged much more than £350 a share – Chief Factors and Chief Traders were allowed to divide among themselves £107,055 as immediate compensation for recent yearly dividends of £275. As before, 40% of the profits would produce future salaries, but the number of shares was increased from 85 to 100. This was done partly to make appointments to two new officer-grades, Inspecting Chief Factors and Junior Chief Factors, and partly to set aside five shares as a retirement benefit fund. Further, the partners were told that they and the Company would share the cost of another new official, a Chief Commissioner "to whom the Officers, Clerks, and Servants shall be responsible."

Unfortunately, there was no guaranteed minimum return in the 1871 Deed Poll, which made 1875 and 1876 particularly dreadful experiences. In those years, no dividends were paid on shares, and subsidies of £100 had to be paid out of the pension fund. (Ultimately, the Company had to guarantee a junior officer a salary of £200 a year, with payments of £300, £400, £500, and £600 to the four higher ranks.) The 1870s and 1880s were grim years of economic depression for many people in North America and in Europe. Falling prices, shrinking markets, and poor harvests caused business slumps in lumber, fish, and wheat. As for the fur trade, competitors were actually outfitting themselves in the Company's new Winnipeg store and then going north to pick off prime furs. Company packers, shippers, and boatmen were demanding higher wages; if they didn't get more money, they sometimes went to work for free traders and competing outfits. Pemmican became scarce as the great buffalo herds of the plains were massacred by hide hunters. But there were even more disturbing factors at work: for various reasons, the fur trade was simply not yielding the spectacular results of earlier decades, and the international fur market was plagued by catastrophic drops in the demand for pelts. In 1883, despite a policy of withholding certain furs from the market in order to support prices, the Hudson's Bay Company reported that the price of Canadian sable was at its lowest in thirty years.

The end of the proprietors in the field was near. In the eyes of the older among them, the fault lay with the avaricious London stockholders. Roderick Ross MacFarlane, a veteran Mackenzie River trader, described the whole situation clearly and concisely, if rather dramatically.

> There is no mistake about it; the fiat has gone forth, and Attila is to ravage and destroy the handiwork of the Company of Adventurers that has reigned in the land for two long centuries and more. The Phillistines [*sic*] are now at last upon us in reality, and there must be a dividend if the heavens should fall . . . exit Hudson's Bay Company; enter Hudson's Bay Lands and Colonization Company Limited.*

The end came in two stages. In Winnipeg in 1887, Chief Commissioner Joseph Wrigley - he whose name had been given to the first steamboat to cross the Arctic Circle and to navigate the Mackenzie to its delta - addressed what turned out to be the last annual council of commissioned officers. He announced that commissions would not be offered to those who had entered the service of the Hudson's Bay Company after the Deed of Surrender that transferred the various territories to Canada in 1869. As a clerk of the time said of the announcement,

> this very startling information threw a wet blanket over the entire service, and produced in many cases very unfavorable results in the Company's interests. . . . It always has been, and is today, conceded by all in the service who are in a position to know that it was one of the biggest mistakes the Company every made when they decided to cut out granting commissions. The only interest the majority of their servants had in the service since then was their weekly, monthly, or annual salary.†

---

* Stanley, G. F. G., "The Fur Trade Party II, United We Stand," *The Beaver,* December, 1953.

† McKenzie, N. M. W. J., *The Men of the Hudson's Bay Company, 1670 A.D.-1920 A.D.,* Fort William, privately printed, 1921.

Then, in 1893, the Deed Poll profit-sharing rights of 1871 were formally cancelled. (In point of fact, the profit-sharing system had ceased to function. For years, a guaranteed income had been an officer's only source of revenue.) The commissioned officers were, officially, as much employees as other HBC personnel. The words "Chief Factor" and "Chief Trader," titles that had been respected, sometimes revered, by generations of Indians, titles that were synonyms for integrity and fair dealing from Labrador to Russian-held Alaska to northern California, now disappeared from the everyday language of Canadians.

* * *

Almost 120 years after its birth in Montreal, the North West Company was dead. But so was its alter ego, the Hudson's Bay Company, because it was upon the system, method, loyalty, and sheer guts of the former that the greatness of the united companies had been based. Each had contributed an equal amount of capital to carry on the business after 1821. More importantly, each had contributed the superlative executive talent without which the trade could never have been carried on so long and so successfully. The organization that, today, refers to itself as "The Bay" has but the slimmest of connections with the Hudson's Bay Company of history. Only the name is the same.

Chapter 5

The Oilman's River

At the time, Alexander Mackenzie was in no position to appreciate it and might well have been politely incredulous had there been someone around capable of explaining the whole thing to him. But, two years before he even set eyes on the river that now bears his name, he recorded a discovery as significant as pioneering a way to Arctic or Pacific tide-water.

The incident occurred within a few days of his first entering subarctic territory. While trudging alongside the ice-strewn Athabasca River, Mackenzie casually noted,

> At about twenty-four miles from the Fork [of the Clearwater and the Athabasca rivers], are some bitumenous fountains into which a pole of twenty feet long may be inserted without the least resistance. The bitumen is in a fluid state, and when mixed with gum, or the resinous substance collected from the spruce tree, serves to gum the canoes. In its heated state, it emits a smell like that of sea coal. The banks of the river, which are there very elevated, discover veins of the same bitumenous quality.*

He had just seen one tiny outflow from subsurface layers of petroleum-soaked ground that stretched around him for several hundred square miles. Today, hardly a month goes past in which we don't read something or other in the newspapers about his "bitumenous fountains." They are the site of two, multi-million-dollar efforts to extract petroleum from what are frequently called the Athabasca "tar sands."

Mackenzie was understandably offhand about an isolated

* *Voyages from Montreal . . .*, London and Edinburgh, 1801.

seepage of bitumen. It was a very well-known substance in his day, its commonest use being as caulking in canoes and sailing ships. Indeed, the human race has made very extensive use of this heavy, almost solid form of petroleum. According to the Old Testament, the ark that Noah built was coated with "pitch within and without," which was undoubtedly bitumen or, to give it a commoner name, asphalt. The walls of Jericho were bonded with bitumen, and the streets of Babylon paved with it. As a medicine, it effectively checked bleeding, aided the healing of wounds, reduced leprous sores, cured chronic coughs and some forms of diarrhoea, and was reputed to be just the thing for straightening out eyelashes. In Mexico, it was utilized as a toothpaste and a chewing gum. In certain parts of South America, bitumen waterproofed the homes of the living and embalmed the bodies of the dead.

With his keen observer's eye and quick intelligence, it's a pity that Mackenzie knew little about geology. Had he studied that science, he might have "read the rocks" and worked out the common factor between the bubbling pools of bitumen next the Athabasca River and certain other geologic phenomena he saw in the course of his wanderings. On his way to the Arctic, there were quite a few of them. He passed by many a limestone outcrop, notably the massive cliffs that form the famous Ramparts, whose walls rear 100-200 feet from the river's edge and for 7 miles confine the wide waters of the Mackenzie to a channel rarely broader than 800 yards. In many places, the river's banks are composed of clay, and near its mouth the land itself on either side is a jumbled mixture of sand, clay, and stone. On the return journey, which necessitated a great deal of wading through the shallows as the men hauled the canoes upstream, Mackenzie had time to observe other signs. He had barely retreated from the enormous, lake-strewn delta when he walked across the mouth of a tributary stream whose banks "tumbled down & broke into small thin Stones like slate but not hard. Amongst the small stones were pieces of petrolium [*sic*] like pieces of Yellow Wax but more friable." Nine days later, when the explorer had back-tracked almost the length of the mainstream, he saw "several Smokes along the Shore

. . . As we approached we found a sulpherous [*sic*] Smell & . . . found that the whole Bank was on Fire for a considerable Distance; that it is a Coal Mine."

On his way to the Pacific, there were still other signs. Twenty-four hours after setting out on his second voyage, he came to a trench-like section of the Peace where the banks had been undermined by the river's spring-swollen waters. "Where the earth has given way, the face of the cliffs discover numerous strata, consisting of reddish earth and small stones, bitumen, and a greyish earth . . ." Later that same month, Mackenzie's second-in-command, after a day's reconnoitring of a mountain trail around the vicious, violent rapids in the Peace River Canyon, reported that he had come across chasms in the earth emitting smoke, heat, and a strong sulphurous stench.

The bitumen, limestone, clay, sand, "petrolium," layers of reddish and grey earth, and burning coal Mackenzie saw all have one thing in common. They are visible reminders of the gigantic, long-gone seas that helped create, among other things, the Canadian prairies. And, since the long valley of the Mackenzie River was moulded by the same natural forces that formed the prairies, it, too, is underlain by deposits of bitumen, oil, and natural gas.

* * *

The word "oil" describes a variety of different liquids used as foods, medicines, perfumes, paints, fuels, and lubricants. Oils were obtained from such organic sources as olives, flax, and the fatty tissues of pigs, cattle, fish, and whales. And there was an oil obtained in minute amounts from the earth and used chiefly as a type of medicine. As early as the fourteenth century, this liquid was commonly referred to as "rock oil." In the Latin favored by the few educated people of that time, it was called *petra* (rock) *oleum* (oil), from which we derive the modern term "petroleum." The name is not as accidental as may appear: petroleum differs in chemical composition from all other oils.

What we commonly call "oil" is a substance varying in appearance from a straw-colored, freely-flowing liquid to a black, sluggishly-moving, almost semi-solid material. No two petroleums are exactly alike, although most petroleums are dark-colored, almost as fluid as water, and composed of liquids in which both gases and solids are dissolved. It is when petroleum reaches the surface of the earth that the gases separate from it, although they sometimes separate underground by natural processes to form natural gas deposits. The solids, too, are sometimes found separate from the liquid components: evaporation of surface seepages of oil leaves the solid constituents behind as one or other form of asphalt, referred to at various times in human history as "burning stone," "slime," "tar," "pitch," and "bitumen." Some of these asphalts are as soft as wax, some as hard as coal.

There's a lot of petroleum in a lot of places. As a matter of fact, after water, petroleum is the most plentiful liquid in the world. Crude, as oilmen call it, is found beneath the fertile farmlands of Holland, the scorched sands of the Sahara, the steamy jungles of New Guinea, and the frozen landscape of the Canadian North. While it is found far underground, you sometimes have to look for it under water, hence the rigs that stand like sentinels in the Persian Gulf, the Gulf of Mexico, the North Sea, and the Beaufort Sea at the mouth of the Mackenzie River.

Although petroleum is found in about fifty different countries, it is by no means easily discovered. It is seldom found where formed. Source rocks, the compacted sediments in which the bodies of marine organisms were, at first, embalmed and then subjected to the pressure and heat that somehow changed their bodies into petroleum, are fine-grained, relatively impervious shales and limestones. Drilling into source rocks produces oil or gas in very little quantity, if at all. For reasons as mysterious as its actual creation, petroleum migrates. Significant quantities are found only in porous, sponge-like sandstones and limestones called reservoir rocks. These strata, whose pores may be so small as to be invisible to the naked eye or large enough to stick a finger into, are what oilmen mean when they talk of "oil pools." There is no actual reservoir of oil or gas: just billions of drops of oil or molecules of gas distributed throughout a particular layer of porous rock – and almost always trapped between non-porous layers. If not trapped, then crude oil eventually seeps to the earth's surface, hence the outflow of "bitumenous fountains" Alexander Mackenzie saw alongside the Athabasca River. Exactly the same is true of natural gas. The Bible's "fiery furnace" in which Shadrach, Meshach, and Abednego walked was very probably a natural-gas leak that had caught fire. The Greeks worshipped at the shrine of the oracle at Delphi, where an eternal flame burned, fed by natural gas leaking from deep rock fissures. On the shores of the Caspian Sea, the Zoroastrians of ancient Persia and other sects who worshipped fire as a god fell on their knees at Baku before perpetual fires, fuelled by natural gas secretly piped from subsurface seepage channels by priests anxious to maintain their positions of prestige and power. Farther east still, the Chinese were quick to find less devious applications for gas. Somewhere between two and three thousand years ago, they tunnelled hundreds of feet below ground and led gas through bamboo pipelines to heat glass retorts and evaporate sea water to produce salt.

Early oilmen in North America, concentrating on surface oil seepages, played hunches. The classic Canadian example is railway-car-maker-turned-oilman James M. Williams of London, Ontario. He bought a controlling interest in a local company owning properties that included Enniskillen Township's notorious "gum beds," several acres of land impregnated by a tarry substance that chemical analysis reported as suitable for making pavements, waterproofing ships, and from which could be processed an "illuminating gas." Here, at the site of what was later christened Oil Springs, Williams built a refinery and processed from these surface deposits of bitumen a foul-smelling, brightly burning lamp fuel that he sold for $1.00 a gallon. Not content with the profits of this operation and suspecting that, somewhere below his feet, there were quantities of free-flowing oil, he started mining the gum beds. When his men had dug through 50 feet of gravel, sure enough they found what Williams was looking for – and North America had its first commercial oil well.* Other oil entrepreneurs relied on guesswork or employed gadgets called "doodlebugs" that were used to discover oil deposits in much the same way that a water diviner uses a forked twig to find water. Most of the time, the doodlebugs didn't work. Most of the time, their users went broke. They were either unaware of, or ignored, the hard fact that you don't look for oil: you look for indications of where oil might be located. To do this, you need to know your geology: considerable knowledge of strata formed by sand, salt, clay, lime – anywhere rock formations suggest the presence at one time of an ancient sea. You also have to have the kind of sophisticated, expensive, research methods and tools that did not become available until the development of twentieth-century technology.

No matter how promising the oil-bearing potential of a given location, only by drilling can you discover if oil actually is there. And drilling is *always* expensive. Permission to work on a piece of land has to be purchased from its owner. Workers, equipment, and supplies, have to be transported to the site, often by air, and maintained for months on end. The oil rig itself, powered by diesel engines and worked by round-the-clock drilling crews, normally costs a cool million – or used to, because inflation has upped that figure considerably. In North America, drilling for oil usually involves going down at least 4,000 feet, so even the steel-toothed drill bit regularly wears out and has to be replaced. Since the bit is at the head of various lengths of drill pipe, every one of these has to be lifted out, unscrewed, stacked, the bit changed, and then all these drill pipes have to be unstacked, screwed back together, and re-inserted into the drill hole. (It takes a seven-man crew all of seven hours to replace a drill bit that wears out 5,000 feet down. And some wells don't yield petroleum until they reach below the 10,000-foot level.) Even in proven oil fields, a new well may turn out to be a "dry" one. In fact, about 25% of all field wells are dry; the percentage of dry holes in new unproven territory – as, for example, the Mackenzie Delta – is up around the 90% mark.

---

* The year in which Williams struck oil is variously given as 1857, 1858, and 1859. However, most authorities seem to agree he did so before, in August 1859, on the south side of Lake Erie, "Colonel" Edwin Drake and William A. Smith brought in a well at Titusville, New York, the name that competes with Oil Springs, Ontario, for the claim to being the first commercial oil well in North America.

Even if oil is found, producing it may not be profitable. The quality of the oil may be too poor or the volume so far below expectation as to be commercially unexploitable. With the exception of exploration in the Arctic, where costs are singularly severe, drilling a wildcat (test well) can cost anywhere from five to ten million dollars – and, as a general rule, only one of every nine wildcats produces oil or gas.

Nothing better illustrates the frustrating nature of the search for large quantities of petroleum than the fact that 90 years elapsed between Williams' finding some oil in Ontario and the discovery of significant quantities of it near Leduc, Alberta, in February, 1947. And this was not for want of looking. At the turn of the century, the West's first real "oil boom" was a small forest of derricks in the Waterton Lakes region of southern Alberta. Here, in 1886, trappers Layfayette French and John George "Kootenai" Brown had learned something from Stoney Indians skimming a blackish-brown, vile-smelling liquid from the surface of a slough at Cameron Creek: it was very useful as machine grease, and also as an external lubricant for minor aches and pains. (A popular use of Waterton oil was as cattle dip to cure mange.) The location was rich in sulphur- and petroleum-laden natural gas, but the know-how and equipment of the times weren't good enough. In the late 1940s, companies were able to exploit these fields properly, using rotary drilling rigs capable of grinding a hole three miles down. Much the same story can be told of Canada's first major petroleum field (that is, a field with at least 100 million barrels* of recoverable gas and oil). It was initially discovered in 1913 in Turner Valley, in the foothills country about 40 miles southwest of Calgary. Even so, for years thereafter, almost every Turner Valley well was a producer of "wet gas," a type of natural gas saturated with condensate, a straw-colored substance similar in composition to a very light-gravity oil. The condensate had to be separated from the gas, which was then wastefully flared at the refinery site (and gave one particular wellhead, whose flame could be seen from Calgary, the proud title of Hell's Half Acre). What these early oilmen didn't realize was that, being lighter than crude oil,

* One barrel of oil equals 35 Imperial gallons.

the gas they were tapping and burning off formed just the cap of a billion-barrel reservoir of petroleum that lay in the pores of a badly tilted limestone layer.

It was the discovery of a major geologic "reef" just south of Edmonton that gave Albertans their first real proof of petroleum wealth – and Imperial Oil had to drill one hundred and thirty-three consecutive dry wells before hitting pay dirt with their one hundred and thirty-fourth. Immortalized in the literature of oil as Leduc Number 1, it made the Canadian West a prime hunting ground for most of the world's major petroleum companies. From a spending rate of something like $1 million a month in 1946, oil investment zoomed to $1 million daily in the decade following Leduc and skyrocketed to $3 million per day in the 1960s. Leduc started a series of explorations that have resulted in the discovery of several hundred oil and gas fields in western Canada. All four provinces have become producers of oil. All, except Manitoba, produce gas. Nowadays, pipelines crisscross the face of western Canada and angle off east, south, and west to supply Canadians and Americans with these fuels and with their many chemical deriatives.

The heartland of the Canadian oil-and-gas industry has long been Alberta, which has been called the "Texas of Canada" with good reason. Its production of fossil fuels has been astronomical: millions of tons of coal, billions of barrels of oil, and trillions of cubic feet of natural gas. Taken together, these amount to something like 80% of all Canada's hydrocarbon fuel reserves found to date. And yet, the reserves known to exist there but yet to be siphoned out of the ground are finite. Turner Valley production has never hit the highs of its pre-World War II days. The wellhead at Leduc No. 1 has given its last drop and has been sealed. And the rate at which we are gobbling up oil and gas – and exporting substantial quantities of them to an energy-hungry United States – has raised considerable doubts as to the ability of the West to continue to meet the demand that will build up, perhaps as early as the mid and late 1980s, but certainly by the year 2000. So where are our supplies of petroleum going to come from? Will we have to pay stiff prices levied by suppliers in Venezuela and the Middle East?

Or is there the possibility of cheaper, "home-grown" supplies in subarctic and Arctic Canada?

* * *

The geologist who was primarily responsible for discovering oil at Leduc in 1947 was the man who, in 1920, produced commercial quantities of petroleum alongside the Mackenzie River barely a hundred miles south of the Arctic Circle. At least, a lot of oilmen credit him with the Leduc find, and one historian of the oil industry describes Theodore August Link as being to western Canadian oil exploration what Babe Ruth is to baseball.

In the spring of 1919, Imperial Oil organized several geological field groups to survey Western Canada from the international boundary to the Arctic Ocean and from the Rocky Mountains all the way east to the edge of the Canadian Shield. Ted Link was assigned the task of checking the lower Mackenzie River region, about 1,500 miles north of Edmonton. The only means of getting there was by canoeing down the rapids-strewn Athabasca River, portaging around sixteen, violent miles of white water on the Slave River between Fort Fitzgerald* and Fort Smith, and later westering across the broad waters of Great Slave Lake to the "top" of the Mackenzie's "trunk". In the summer of that year, Link and his crew encamped 50 miles below the HBC post of Fort Norman. Despite having to fight off mosquitoes "big enough to shoot with a rifle" and run a battery of tests on subsurface soil that was permanently frozen, they became quite convinced that they'd found, in their leader's words, "a well of undoubted commercial possibility." By late August, a six-man drilling team and equipment had arrived to undertake the slow, monotonous job of excavating a bore hole with a cable tool rig, a primitive form of drilling that punched a

---

* Until 1911, this tiny settlement was known as Smith's Landing. That year, Sergeant F. J. Fitzgerald of the Royal North West Mounted Police perished in the course of a patrol from Fort McPherson on the Mackenzie River to Dawson, then the capital of the Yukon Territory. Pioneers up and down the river greatly mourned the death of this fine man and, in order to perpetuate his name, a group of them decided to give it to the hamlet at the head of the rapids.

hole in the earth by repeatedly lifting and dropping a heavy cutting tool suspended from the end of a cable. All of which was just the beginnings of a lot of trouble.

Link stayed on long enough to hire an ox from the HBC to do the heavy work of hauling logs into camp for use as winter fuel and then left for Edmonton to plan the next summer's activities. In the spring, he returned the hard way, helping to pole a 40-foot scow heavily loaded with additional equipment, and staying aboard the emptied scow when it was run through the notoriously murderous rapids on the Slave River. On the Mackenzie mainstream, time and time again he labored alongside the scowmen as they unloaded the entire cargo in order to work their vessel clear of a sandbar and then manhandled everything back on board again. Upon arriving at the camp, Link discovered that some members of the drilling team who had wintered there showed signs of being badly "bushed." After working in temperatures of 40-60 below, they tired easily and were so apathetic that they had to be constantly chivvied to perform even the most routine chores; they were inclined to sit around daydreaming to the point where they were hallucinating. Three of them became so moody and despondent that Link was forced to send them "out" to recuperate.

Drilling had barely reached the 400-foot level and, with only three experienced well-head men left, progressed slowly. In addition, they had to do any lifting or hauling of heavy loads using their own muscle power because the ox had made an excellent series of steak and stew dinners when its hay supply had run out. So Link had to content himself surveying and mapping the Mackenzie and the surrounding neighborhood. And it was while away from camp in the late summer that he got the glad news. Alf Patrick, the head driller, came in search of Link to tell him that, after getting to the 700-foot level, oil was bubbling over the casing at the bottom of the derrick. His boss calmly nodded acknowledgement and replied, "Call me when it tops the derrick." Off went a subdued Patrick, only to come pounding back later that day yelling, "It's blowing all over the place!" Link returned to the well in time to witness oil shooting 60 or 70 feet into the air.

Link was pleased. He'd hoped for a big producer and he'd got it. In fact, had the *Guinness Book of Records* existed at the time, Ted Link would have rated an entry because he had located the first commercial quantities of crude oil in western Canada. (Greater quantities of crude were not found until 1936, on the west flank of Alberta's Turner Valley.) He immediately began buying land claims from local Indians, paying them in cash from a bag in which he carried $20,000 in one-dollar bills. (Only one landowner refused the money – he preferred to be given a new canoe.)

But, if Ted Link had raised oil, he'd also raised a problem. How was he going to get it to market from Norman Wells, where his men went on to drill several more successful bore holes? Local markets didn't exist, and Edmonton, the gateway to southern Canada, was 1,500 miles to the south by water and wilderness trail.

It was Charles Taylor, the western development manager for Imperial Oil, who came up with an answer. As the crow flies, Norman Wells is a mere 900 miles north and west of Edmonton. So why not use airplanes to fly in men and materials to stake further claims and bring out oil? As Taylor informed head office in Toronto, Edmonton had both airfields and unemployed pilots.

His brainstorm was approved, and enquiries were sent out to various plane manufacturers and suppliers. In the end, two First-World-War-vintage Junkers aircraft were purchased by Imperial Oil from an American aviation firm and ferried in slow stages from Long Island, N.Y., to Edmonton, Alta. These "huge, corrugated monsters," as newspaper reports described them, were all-metal, all-weather monoplanes. Each was powered by a single, 175 hp. engine and equipped with an undercarriage so designed that it could be fitted with pontoons or skis. If any aircraft of that time could beat adverse weather conditions, a Junkers could. And the huge aircraft seemed to prove this in the course of the month-long, but accident-free, haul from Long Island in the winter of 1920-21, when a combination of blizzards and icy fogs made any form of prairie travel extremely hazardous. Oddly enough, the end of the next northerly leg was a curious crossing of historic trails. The first aircraft bound for the

Northwest Territories touched down at a hastily-built air base close to the junction of the Peace and Smoky rivers, a few hundred yards from the site of the fur post where the first explorer of the Mackenzie River wintered in 1792-1793 before making his overland dash to the Pacific.

"Rene" and "Vic," as the Junkers had been christened, did not arrive at Norman Wells as anticipated – by air. After surviving a four-day battle with strong headwinds and heavy snowstorms, they managed to reach the HBC depot of Fort Simpson. Here, while putting down on the runway of the Mackenzie River, "Rene" hit an ice barrier in the form of frozen dog-sled tracks and upended. When the huge cloud of powdery snow thrown up by the crash cleared, pilot George Gorman climbed out of the cockpit to find the undercarriage twisted and the propeller broken. "Vic" landed safely, although its engine was now functioning erratically due to a build-up of carbon. Some fast repair work and the loan of "Vic's" propeller and one of its skis made the big Junkers airworthy again. But something went wrong on takeoff. "Rene" fell fifty feet to the river surface, crushing the undercarriage, shattering the borrowed propeller, and demolishing one wing.

The mechanics who flew with bush pilots performed small miracles to keep an aircraft serviceable, but Bill Hill and Pete Derbyshire reached a new high in ingenuity – and created a northern legend – when they got "Vic" back into the air. Cannibalizing engine parts from "Rene" was easy. However, what do you do when you lack a delicately engineered mechanism such as a propeller? They couldn't order one in because there was no radio at Fort Simpson, and the nearest telegraph was 1,000 miles away. So they made one on the spot. In fact, they ended up making two. With help from Walter Johnston, an HBC steamboat engineer and former cabinet-maker, some of the other inhabitants of Fort Simpson, notably Father Decoux of the Roman Catholic Mission there, Hill and Derbyshire did exactly this. Somehow, they fashioned aerodynamically-sound propellers out of oak sleigh boards, some large metal clamps, and several potfuls of glue made from moose hide. These unorthodox replacements worked beautifully – and, in the case of "Vic," in the nick of time. The ice on the Mackenzie broke up early that year. It was heaving and grinding loose as Elmer Fullerton eased his huge transport gently off the river. According to one eyewitness, as "Vic" struggled to gain altitude, the heels of the skis momentarily trailed through a large pool of meltwater.

Re-equipped at Peace River Crossing (the modern town of Peace River) with a factory-made propeller and carrying various replacement parts for "Rene," sitting beached at Fort Simpson, "Vic" winged north again, made its deliveries, and carried on downriver to the HBC post of Fort Norman. There, Fullerton brought the Junkers gliding down on to the glass-smooth Mackenzie. Just before "Vic" came to a stop, the right wing started to drop: a pontoon had been ripped open by a "dead-head," a water-logged tree floating just below the surface. So, Fullerton had to taxi the last fifty miles downstream to Norman Wells, one wing supported by a barge in order to keep his aircraft on a reasonably even keel, an inglorious and, as it turned out, a temporary end to any attempt to establish air communication with the far north. No more airplanes attempted to penetrate the Arctic for several years.

Ted Link, who had been aboard "Vic" all the way from Peace River Crossing, shrugged off the failure to establish air links with the "outside." He had another, bigger worry. His men had managed to distill a semi-refined gasoline from Norman Wells' crude to aid "Vic's" return to Peace River Crossing (where the ever-unlucky "Rene" had a hole torn in a pontoon, flipped over onto its back, and ended up more or less a wreck). But the airmen were the only available customers of any consequence. Even then, Link was "selling" about 200 gallons of gas to Imperial Oil. HBC sternwheelers on the river burned wood, as did local residents. Link's problem was a marketing difficulty that neither he nor his employers could solve in the 1920s. Events elsewhere did. When large deposits of uranium were found nearby on the shores of Great Bear Lake in the early 1930s, Imperial Oil was able to uncap the six wells then drilled and operate a 500-barrel-a-day refinery at Norman Wells to service the needs of the miners at Port Radium; two years later, the discovery of the Yellowknife gold field on the north shore of Great Slave Lake gradually upped capacity to 840 barrels a day.

Further expansion at "the Wells" came as the result of World War II. In late 1941, Japanese forces occupied the Aleutians, the chain of islands off southern Alaska that is a series of stepping stones leading to the heart of the North American continent. In response, the Canadian and American governments worked out a complex plan to build a 600-mile pipeline westward from Norman Wells to a refinery (imported at horrendous expense from Texas) at Whitehorse, Yukon Territory. From there, gasoline and fuel oil would be forwarded to airfields and encampments via the Alaska Highway that was being constructed by American and Canadian army engineers. All this is another story, and there's no space to tell it here. But one result was a contract whereby, for a $1.00 a day and expenses, Imperial Oil sent personnel to Norman Wells in 1943 to try and supply the United States Army with 1,000 barrels of oil a day. These men succeeded in drilling sixty extra wells, which gave the Canol* pipeline a capacity of 4,000 barrels a day. In fact, by the end of 1945, something like one million barrels of oil had been pumped over the Mackenzie Mountains to help turn back the threat of Japanese invasion.

All of which caused someone to remark that the pace of northern development was obviously set by the market place, not by physical and engineering difficulties. The facts back up the remark. Among other things, the Canol Project involved building a series of airfields from Alberta to Alaska, constructing the Alaska Highway, and then laying a small-diameter pipeline across the thickly forested Mackenzie valley, up and down the canyon-strewn flanks of the Mackenzie Mountains, and over the tundra interior of the Yukon Territory. It took nine pumping stations to keep the oil moving and raise it to the line's highest point at Mile 86: 5,860 feet. Add the wages and provisions of thousands of military personnel, the cost of storage depots, road-building materials, engineering equipment, airplanes, and so forth, and the final bill was somewhere around 134 million

* A contraction of Canadian Oil Project.

dollars – and all this effort and money was for a one-year supply of petroleum!

* * *

The drainage basin of the Mackenzie River contains two, potentially enormous, sources of petroleum. One is located at its mouth, the other well upstream. Of the two, the latter may be the greater source – if the oil industry can come up with the technology that will do the job of extracting petroleum efficiently and cheaply. At the moment, this is one of the big "ifs" of human history, and its story begins, just as this chapter did, with the pools of bitumen Alexander Mackenzie saw alongside the Athabasca River.

About twenty-five miles upriver from the place where Mackenzie caught his first glimpse of "bitumenous fountains" is the junction of the Athabasca and Clearwater Rivers. Here, Mackenzie's fur-trade colleague and predecessor, Peter Pond, had built what he called the "Fort of the Forks." In the course of the next hundred years, no less than five fur posts were operated at one time or other by Nor'Wester or Bay men at this river junction. In 1870, HBC man Henry John Moberly established Fort McMurray on the site of one of these posts, not for fur storage but as the future terminus of a steamboat route that would extend down the Athabasca and Slave rivers as far as the rapids halfway along the latter and, ultimately, all the way to the Arctic Ocean. However, for all that it became an important trans-shipment point for scow and steamship loads of HBC goods going north and fur bales coming south, Fort McMurray was no metropolis. It was described in 1913 by Sydney C. Ells, a young surveyor, as

> . . . consisting of a dozen primitive log cabins, a bug-infested hovel proudly referred to as a "hotel" and, during the summer months, many Indian teepees and tents. Everywhere starving train dogs roamed at will, and the greatest care for the protection of food and other supplies was essential. . . . The community depended solely upon the fur trade . . . Damage to, and loss of, goods in transit dictated commodity prices, and salt and sugar . . . sold at three pounds for a dollar. Consequently

McMurray was known as one of the "three-for-a-dollar settlements."*

In 1908, Inspector W. H. Routledge of the Royal North West Mounted Police set off from Edmonton on one of his routine northern patrols. He received a couple of surprises along the way. So many free traders were competing with the HBC that some Indian cabins now boasted gramophones, musical clocks, sewing machines, and Huntley and Palmer's best tinned pound cake, items considered too frivolous for HBC stores of the time. Another change was the skyline north of Fort McMurray, against which reared the stark skeletons of several, handhewn oil derricks. Various individuals were hard at work trying to tap the Athabasca tar sands, spurred on by the reports of turn-of-the-century federal government geologists that they contained "evidence of an upwelling of petroleum unequalled in the world."† However, local interest in the tar sands was non-existent until about five years later, when the surveyor who described Fort McMurray as a "three-for-a-dollar" settlement mapped the largely unknown deposits and brought out the first large quantities of the strange sands for study. Sydney C. Ells, an engineer by trade, became so obsessed by the sands that he spent the next thirty-two years of his life studying them, lecturing about them, and attempting to persuade industry and government to get together to exploit them. To prove that his judgment of the sands' utility had point, Ells hauled sixty tons of the stuff out of Athabasca during the bitter winter of 1914-1915 and paved part of an Edmonton street with their bitumen. Ells proved his point – but for one thing. It was much cheaper to import asphalt paving from elsewhere.

---

* Ells, S. C., "Recollections of the Development of the Athabasca Oil Sands," Ottawa, Information Circular 139, Mines Branch, Department of Mines and Technical Surveys, 1962.

† Strictly speaking, the term "tar sands" is inaccurate. Tar is a residue of a particular petroleum-refining process. The sands are bitumenous, that is, their chemical composition is that of a hydrocarbon in raw form, as is that of any petroleum deposit. But, like any nickname, the term "tar sands" sticks.

Expectations prompted by Ells' investigations started the first real boom in Fort McMurray, and several homesteaders suddenly found that they could sell their property for thousands of dollars. This signalled the start of a different history for the town in the twentieth century, which can be aptly summed up as "boom and bust, bust and boom." The first "bust," the depressing discovery that the tar sands were not at all an easy economic proposition, killed real-estate speculation and shrank offers of $1,500 an acre to $8.00. For the first, but not the last time, furs, lumber, and fishing again became the major occupations of Fort McMurray residents. In the words of one writer, the town "shrank into a pocket of poverty in the midst of the affluence of oil-rich Alberta . . . The people of Fort McMurray no longer cared. They had long since lost all faith and hope in the plans of oil men. They knew that literally beneath their feet lay the largest known deposit of oil in the world, and for all that they could tell it would be there forever."*

There were, however, several men who were determined that all this petroleum would not stay locked in the sands of Athabasca. And their efforts make a curious, still not fully known, tale of hopes realized and hopes deferred – and of government stupidity or duplicity. But first, a word or two of explanation about the strange, sticky sands themselves.

Imagine a sandbox roughly the size of the province of New Brunswick into which at least 600 billion barrels of bitumen have been poured, mixed with generous helpings of clay and assorted minerals, and then topped off with a crust of glacial stone and rock anywhere from 150 to 2,000 feet deep. Cut a piece out of this gargantuan, muskeg-sprinkled "pie" and the "filling" is a black, gooey, sulphurous-smelling dirt. This so-called "tar" is bitumen, a very heavy form of crude oil which, even in a liquid state, is about as lively as the thickest of molasses. Heat a glass tumbler, fill it with bitumen, turn the glass upside down, and the bitumen displays the slowest of slow-motion movement. At room temperature, it stays put in the tumbler. It is for this reason that bitumen cannot be pumped through a pipeline, or even easily handled, unless

* Gray, Earle, *The Great Canadian Oil Patch*, Toronto, Maclean-Hunter Limited, 1970.

diluted with a solvent or kept in a heated condition. In appearance and texture, this gummy dirt is for all the world like asphalt paving, although only about 12% of it, by weight, is raw bitumen.

Just how this variant of petroleum was formed and why the only major Canadian deposits are in Alberta have long puzzled scientists.* Some say that this particular form of petroleum was, at one time, a lighter oil that migrated from other strata to its present locations and was then converted into a tarry substance; others speculate that it is really a sort of proto-petroleum, which is still in the early stages of developing into the chemically lighter and more familiar forms of crude. Whatever its origin and history, this bitumen does have two distinct disadvantages in the eyes of oilmen. It has a much lower ratio of hydrogen to carbon than the conventional crude found by drilling, yet the higher the ratio of hydrogen to carbon, the higher the energy of a fuel. Second, it has a high sulphur content, an undesirable constituent for various reasons. Nonetheless, raw bitumen can be processed and upgraded into a desirable synthetic crude – if it is separated from all the other ingredients of tar sands.

The first man to succeed in producing crude oil from tar sands on a commercial basis was a Maritimer, R. C. Fitzsimmons. Attracted to Fort McMurray in the early 1920s by rumors of the oil wealth of the sands, his entrepreneurial instinct was encouraged by what he saw there. He bought about six thousand acres of tar-soaked land some fifty miles downstream from the town, a holding he named Bitumount.

---

* The Athabasca tar sands are far from unique. They are the *largest* known deposits of bitumenous sands in Alberta, but not the only significant ones in that province, in Canada, or, for that matter, in the world. West of the Athabasca deposits are the Peace River oil sands (estimated to contain approximately 51 billion barrels of bitumen), south and west are those of Wabasca (54-billion-barrel estimate), and to the south-east the Cold Lake deposits (164-billion-barrel estimate). All these differ from the Athabasca sands in that they are deposits located too deep to be recovered by existing mining methods, although the industry is desperately trying to find a means. Widely separated outcrops of oil-soaked sands have also been found on the north-west tip of Melville Island in the Arctic. And there are massive oil-sands deposits in Venezuela (approx. 200 billion barrels) and on the island of Malagasy (approx. 2 billion barrels) in the Indian Ocean.

Drilling failed to uncover the "pool of free oil" everyone said saturated the tar sands. Attempts at *in situ* (in place) extraction by heating underground sand layers with steam and pumping bitumen to the surface also failed. So Fitzsimmons set up a plant to produce crude by literally boiling the tar sands in water: sand settled on the bottom, while a sort of very crude oil frothed on the top, where it could be skimmed off. By 1930, Fitzsimmons' International Bitumen Company (IBC), a federally-incorporated company, was offering several hundred barrels of a tarry substance that was promoted with true advertising panache in western Canada. This product, Fitzsimmons claimed in his advertising, was excellent for "paving, laying built-up roofs, processing into roof coatings, plastic gums, lap cement, caulking compounds, waterproofing, marine gum, fence-post preserver, boat pitch, mineral rubber, and skin-disease medicine." Sure enough, a Calgary company did succeed in selling a number of such products made from IBC's crude. And a few miles of road in Banff and in Medicine Hat were paved with asphalt derived from IBC crude. But purchasers fed up with processing Bitumount crude differently from conventional supplies of crude oil forced Fitzsimmons into financing a refinery that would break down the bitumen into what his customers called "specification products."

When trying to raise capital from time to time to expand his operations and derive more products from the bitumen, Fitzsimmons kept running into some very peculiar difficulties. On one occasion, a share issue offered in Montreal nearly raised the necessary amount of money, but failed when the Alberta government distributed a description of tar-sands development that made no mention of the International Bitumen Company, then the only commercial petroleum-extraction outfit in the region. Attempts to raise the money in London, England, and in Chicago failed because, according to Fitzsimmons,* the potential money-lenders could never get the Alberta government to substantiate the existence of the IBC operation. However, raising a few dollars here and a few dollars there, Fitzsimmons kept

* Fitzsimmons, R. C., *The Truth about the Alberta Tar Sands: Why They Were Kept Out of Production*, Edmonton, privately printed, 1953.

his plant going. By 1938, it was turning out annual production runs of 4,500 barrels of specification asphalt and 2,000 barrels of fuel oil that met industry standards. Yet marketing was still difficult, this time because, again according to Fitzsimmons, customers reported they had been warned by "conventional oil suppliers" that if they bought IBC products, they would be cut off from these same conventional supplies.

Fitzsimmons stayed in business by the simple expedient of borrowing more money. By 1942, with a war going on, asphalt selling at $12 a barrel and his plant putting out 500 barrels a shift, he would be free of debt in no time at all if he could get a large loan. He applied to the province for $50,000, asking that this be given either in the form of a loan or a prepayment for asphalt supplies. But Fitzsimmons' application was refused by the government led by Premier William Aberhart, the fiery apostle of – would you believe – Social Credit.

In his book *Fuelling Canada's Future* (Macmillan of Canada, 1974), Wade Rowland remarks that the International Bitumen Company was ripe for a takeover by someone who had the contacts Fitzsimmons lacked. Rowland records that "into the company's offices that fall strode Mr. L. R. Champion of St. James Street, Montreal, a man unknown to Fitzsimmons but well known to the Alberta cabinet and to its minister of lands and minerals, Nathan E. Tanner. Champion offered to take over the company and put it on a sound financial footing, and after protracted negotiations, Fitzsimmons accepted the offer." A new company, Oil Sands Limited, was formed and took over not only the assets and liabilities of IBC but the patent on its method of oil extraction. In 1944, Champion incorporated yet another company and used all its possessions as collateral to obtain from the Alberta government provincial monies with which to build a brand-new extraction plant. When Champion did not meet the terms of this agreement, the Alberta government took over everything – including the valuable patent on oil extraction. By 1950, that government was informing the North American oil industry of Alberta's wish to see the oil sands developed and was inviting offers.

Bob Fitzsimmons wasn't the only one to suffer setbacks in

early efforts to exploit the tar sands. At the urging of that indefatigable tar-sands enthusiast, Sidney Ells, American oil magnate Max W. Ball organized Abasand Oils Limited (AOL) in 1936. Four years later, using a hot-water extraction method similar to the one being used by Fitzsimmons, AOL's plant near Fort McMurray was processing several hundred tons of sand a day. By 1941, annual production was 17,000 barrels of bitumen, which was refined into gasoline, diesel fuel, and home-heating oil. But that same year, the plant was destroyed by fire. Worried by wartime shortages of petroleum, Ottawa stepped in and took over the Abasand operation. Work began on reconstructing the plant, but, before this was completed, another devastating fire occurred. Strange stories of incompetence, misuse of public funds, and sabotage began to drift south from Fort McMurray, and the Alberta government tried – without success – to find out what had been going on at the junction of the Athabasca and Clearwater rivers. After all, this was provincial territory, and since 1930 the province had owned sole rights to all mineral deposits found anywhere in Alberta. Why then had the federal government held on to 2,000 acres in the McMurray area, ostensibly for paving roads in national parks? Was it true that, at a cost of $2 million, federal surveyors had found a very rich body of bituminous sands at Mildred and Ruth Lakes, twenty miles downriver from Fort McMurray? Was the government of Canada trying to muscle in on a valuable provincial resource? And why did Ottawa refuse to appoint a Royal Commission to investigate the second fire at the Abasand plant, where newly-installed water pumps didn't function as the place burned down?

The thorny issue of provincial rights has long made Canadian politics a prickly profession. Federal-provincial conflicts have ripped parties apart and ruined many a promising political career. And there are few, more bitterly contested arguments today than who deserves the taxes levied on the extraction and sale of petroleum in Canada's western provinces. This issue goes all the way back to the last year of World War II, when the Social Credit government of Alberta premier Ernest Manning accused the Liberal government of Prime Minister William Lyon Mackenzie King of a "wanton plunder of provincial rights" and discredited the feasibility of exploiting the tar sands.

Thereafter, a lot of interesting developments took place within a very short space of time.

In 1948, the Manning government took over control of Champion's Oil Sands Limited and the extraction plant at Bitumount, which used the patented extraction method that had originally been developed by Bob Fitzsimmons. The intention was to demonstrate to the petroleum industry that development of the tar sands was a commercial proposition. In 1949, under the leadership of a professor of metallurgy at the University of Alberta, Dr. Karl A. Clark, a long-time friend of Ells and a long-time investigator of hot-water methods of extracting bitumen from the sands, the Alberta Research Council operated the plant as a pilot project. A year later, Clark reported that 90% of the bitumen had been recovered from the processed sands and that the way was open for study of "a complete sequence of operations, from mining to marketing." Whereupon, the plant was promptly closed down.

Next, the province appointed Clark's former student and long-time assistant, Sidney M. Blair, a founder of the engineering firm of Canadian Bechtel (a subsidiary of the giant Bechtel Corporation of San Francisco, then interested in building pipelines from the tar sands), to examine the economics of tar-sands development. In December, 1950, the Blair Report outlined a 20,000 barrel-a-day, bitumen-extraction operation in the Mildred-Ruth Lakes area that would produce upgraded synthetic petroleum, deliver it to the American Mid-West, and still give a return of 5½% on investment. The Report emphasized that the major ingredient of successful production was large-scale operation. In fact, Blair proposed what was, in those days, a terrifying large-scale extraction-cum-processing operation: a practical plant would need, he said, not an investment of Abasand's $500,000 but something of the order of $250 million.

In the fall of 1951, in the Alberta capital of Edmonton, the government of Premier Ernest Manning invited the petroleum industry to attend an "Oil Sands Conference." In the course of this get-together, the executives of the world's leading petroleum companies were given technical information on the physical nature of the tar sands, the location of known deposits, the various extraction techniques developed to separate the raw bitumen from the stubborn sands, and even the means whereby this bitumen could be upgraded to produce petroleum. Presentations were made of Blair Report statistics. Then an additional incentive was introduced: a provincial leasing policy that was irresistible. Provided an applicant fulfilled certain, more or less minor, prospecting conditions, exclusive rights would be granted to develop the oil sands (in blocks of 50,000 acres, one block per applicant) for a 21-year period, *renewable for a further 21-year term* for an annual fee of $1.00 per acre (later reduced to 25 cents). Further, royalties charged on any development would not exceed 10% of the value of the raw bitumen extracted. By comparison, the leasing of petroleum deposits in the oil shales of Colorado, Utah, and Wyoming, where extraction problems are even more difficult than those in the tar sands, starts out at 25 cents an acre in the first year and goes to 50 cents thereafter, *plus royalties based on the grade of shale being mined.* The winning bid on one of the first public tracts of oil shales to be leased was $42,000 an acre!* (Unlike the Alberta tar sands, the development of these oil-shale deposits is being allowed to proceed very slowly and only under careful official observation. As environmentalists point out, large-scale shale mining will lead to air and water pollution, and a tremendous upset of natural balances as masses of people and machinery move into areas previously undisturbed by human settlement and activity.)

The Manning government doubtless had the future interests of its citizens at heart. Nonetheless, the 1951 Oil Sands Conference was an invitation to commit theft. And the invitation was so naked, that this is exactly what the petroleum industry did. Most of the prime acreage was snapped up right away by major oil companies, a few of whom now own the rights to petroleum deposits estimated to be about 25% of the world's known oil reserves.

Somewhat ironically, the first company to initiate large-

* Pratt, Larry, *The Tar Sands: Syncrude and the Politics of Oil*, Edmonton, Hurtig Publishers, 1976.

scale development of the tar sands was a direct descendant of International Bitumen Company. It's a complicated, corporate story, but what started out as Bob Fitzsimmon's hand-to-mouth outfit operates today under the name of Great Canadian Oil Sands (GCOS), a subsidiary of the Sun Oil Company of Philadelphia (known to you and me as the local "I-can-be-very-friendly" Sunoco dealer).

The official opening of GCOS occurred in September 1967, when several hundred people gathered on the west bank of the Athabasca River, some twenty miles north of Fort McMurray by dusty, gravel highway, to witness the opening of what has since been billed as the "world's first oil mine." One of the more important guests was the slim, austere-looking, serious-minded premier of Alberta, Ernest Manning. Another was the equally conservative, patriarchal, and religious-minded John Howard Pew, the eighty-five-year old chairman of the board of Sun Oil. However, it was one of Manning's cabinet colleagues who had done the spadework that induced John Howard Pew to turn up in the bleak wilderness of tamarack, spruce, and pine edging the Athabasca River. Three years earlier, Russell Patrick, Alberta's minister of mines and minerals, had been the only outsider in attendance at a Sun Oil Company directors' meeting in Philadelphia when the decision was made to continue the oil-sands development begun in 1963. It was a bold decision. Early GCOS experimentation, followed by a non-confidence attitude among important investment brokers, strongly suggested that any effort to get petroleum from the sands was a gamble against unbeatable odds. In fact, GCOS officials were unable to recommend to their bosses how the project could be kept alive. And then Patrick made his pitch for going on, aiming his challenging comments at the chairman of Sun Oil.

> When Patrick had finished, all eyes turned towards the head of the table to watch the old man rise with a new gleam in his eye as he set off to tilt with the greatest challenge in his career.
>
> "Young fellow," he said, looking at Patrick, who was never to forget his words, "you've caught me in a good mood. Today is my eighty-second birthday, and this morning on the course my golf

score equaled my age. For many of those years I have met many a challenge, and at times lost, so I'm glad you didn't come here to try to soft-sell me, because that doesn't work.

"As I see the situation, perhaps Sun Oil has been remiss in its exploration program aimed at securing ample future supplies of oil for its integrated company operation and in not taking a more significant share in the sands and a share in their risks. Personally, the sands have always fascinated me but I have been so busy."

Then, after a pause, he continued: "And yet the years are catching up on me. When the world looks back it will soon forget that I built up the Sun Oil and put together my large shipping interests – maybe, however, it will not forget the man who provided the key to unlock the riddle of the oil sands – we'll go for it."

Having appalled his colleagues with that pronouncement, he starting acting upon his decision immediately and, turning to his nephew and successor, he said: "That 10,000 barrels per day we get from Venezuela, we will replace that with oil from GCOS. And we'll take other measures."

Then, coming around the table to shake hands with Patrick, he repeated: "We'll go for it."*

It was easier said than done. On the 4,000 acres of Mildred-Ruth Lakes Lease No. 86, the bed of oil sand is buried under a relatively thin layer of glacial overburden averaging 50 feet in depth. The production of crude from bitumen begins with open-pit mining: conventional earth-moving equipment strips away the overburden of muskeg, rock, and clay. A relatively easy job in summer, it becomes a nightmare task in the sub-zero winter weather of the Athabasca country. Metal becomes brittle, and diesel engines freeze up. The buckets of the excavators grow red-hot and gradually buckle as they scrape out the cement-like tar sands. To move the sands to the separation plant is something else again. GCOS uses gigantic bucket-wheel excavators, each of which gouges about 100,000 tons out of the tar-sand layer per day – the equivalent of a railway train of gondola cars 16 miles long – and dumps these onto several

* MacGregor, J. G., *Paddle Wheels to Bucket-Wheels on the Athabasca*, Toronto, McClelland and Stewart Limited, 1974.

miles of conveyor belt. Enormous amounts of water have to be pumped to the separation plant. (Equally enormous amounts of tailings, water laden with sand and chemicals, are later channelled from the plant into settling ponds.) The complicated hot-water flotation process that separates the bitumen from the sand produces a tarry scum that has to undergo considerable further treatment. At this point, the bitumen is still too viscous to be of any use, so it is pumped into huge coking drums, where the high sulphur content is removed. Then a hydrogenation treatment upgrades the bitumen to a synthetic petroleum comprised of gasoline, kerosene, naptha, butanes and pentanes, and an oil somewhat heavier than furnace oil.

Why go to all this trouble and incur all manner of expenses to end up with 65,000 barrels a day (b/d) of petroleum when the daily consumption in the United States alone is around 18 million b/d? GCOS shareholders have watched stock they bought for about $14 hit an early, all-time high of $17 and then steadily decline. In 1977, an ordinary share was worth about $5, and common stock of GCOS has yet to pay a dividend. After seven years of million-dollar costs that often far exceeded budget, the GCOS strip-mining operation finally showed a tiny profit in the second quarter of 1974. But this was a quite nominal profit. It was only made possible by the Alberta government's agreeing to forego $7 million in royalties in the period 1967-1974, another $6 million in sales taxes being waived by the federal government. So why is oil being "mined" in Athabasca?

One reason is the chemical nature of bitumen-derived petroleum. After bitumenous sands have been coked, desulphurized, and hydrogenated, the resulting product is a petroleum of generally higher quality than that of conventional well-head crude. True, several tons of material must be mined, moved, and processed to obtain a barrel of what the industry calls "synthetic crude oil." However, as a petrochemical "feedstock," this particular brand of petroleum is valuable. All petroleums have one thing in common: impurities apart, they are composed of varying proportions of carbon and hydrogen combined into compounds known as hydrocarbons. The chemically "smallest" hydrocarbon compounds make up natural gas, the largest, solid asphalts. Some are the right chemical "size" for refining into solvents and gasolines, or into kerosenes, or lubricating oils, or waxes, plastics, paints, pharmaceutical products, etc. Now, each oilfield produces a different kind of crude petroleum. Each crude has to be analysed and evaluated as a separate entity. Indeed, the almost infinite variation in the chemical composition of petroleum crudes makes their classification very difficult. However, as a general rule, the higher the gasoline content of crude, the higher its worth to a refinery. Crude oils with high gasoline contents have high API gravities, which are ratings from 0-75 on a scale known as the American Petroleum Institute or API scale. Thus, a report that a field such as Leduc was producing a 39 gravity crude brought big smiles to the faces of the drillers and petrochemical engineers alike. The crude being turned out by GCOS has an API rating of around 42.

A second reason why oil is being "mined" is that there's a fantastic amount of it lying around in northern Alberta. Exactly how much petroleum there is in the province's major tar sands deposits* is not known, but even an average of most estimates sounds incredible. The Athabasca tar sands alone are thought to contain 600 billion barrels of petroleum, which is approximately 25% of the supply of oil known to be available from conventional sources around the world.

* * *

It is in one particular section of the immense Athabasca tar sands deposit that all the action is presently taking place. Just north of Fort McMurray, in a long, narrow strip of territory bisected by the Athabasca River, lie rich, thick deposits of tar sands. Here, perhaps 25-30 billion barrels of

* Where tar sands are too deeply buried by overburden to be mined, *in situ* (in place) methods are being developed. Currently this involves heating or otherwise dissolving the heavy, sluggish bitumen and then pumping it to the surface by conventional oil-well methods. A great deal of time, energy, and money is being spent on *in situ* methods by Imperial Oil, Shell Oil, Texaco, Amoco (Standard of Indiana), and Petro-Canada, the company set up in 1975 by the federal government to engage in oil-and-gas exploration and development. But, so far, nobody has made any money at it.

petroleum can be recovered although, by much painful, costly experiment, GCOS has only managed to extract a few million barrels from its particular lease alongside the Athabasca. The highway from Fort McMurray runs along the western edge of GCOS' property. Just down the road on the opposite side are the major facilities, construction sheds, and workers' trailer camps of the Syncrude Project, the only other petroleum-extraction venture in the region.

Visit Syncrude Lease 17 and you get a funny feeling you're peering into the future. Everything is done on an even more stupendous scale than the GCOS operation. Seven thousand acres of muskeg are being drained. The layers of soil above the tar sands are being systematically peeled off by teams of bulldozers and dump trucks so large that their wheels dwarf the men and women standing alongside them. Other workers have created lake-like tailing ponds, waste-disposal lagoons in which will be dumped the oily, chemical-laden residues left over from the extraction process. Still others construct the huge, silvery-grey extraction plant, or work on the slim, rocket-shaped towers of the refinery. Gigantic cranes, each equipped with a bucket capable of holding 100 tons of material, tear monstrous trenches out of a sombre-black landscape fringed by the dull-green subarctic forest. A small army of electricians, welders, pipefitters, and other workers is preparing the means whereby an estimated 125,000 barrels of high-grade petroleum per day will be pumped through Syncrude's own pipeline to petrochemical plants in southern Canada and the northern United States.

The cost of all this? It's unbelievable, or rather incomprehensible. You have to add up the wages of unskilled laborers making, with overtime, anywhere up to $18,000 a year, pipefitters and boilermakers earning up to $35,000, free board and lodging for some 6,000 workers (and beer sold on site at the "Muskeg Pub" at a subsidized rate of 25 cents a glass), huge imports of machines and machine parts, massive purchases of steel, and the construction of a 275-mile pipeline to Edmonton, because the only other pipeline belongs to GCOS. In the end, the total is somewhere around $2 billion. That's the equivalent of every Canadian man, woman, and child putting $1,000 into the same kitty. As a matter of fact, Canadian taxpayers are doing just that – contributing money to the Syncrude treasure chest. And the story of how this money is being raised and spent is much more peculiar than the tale of Bob Fitzsimmons' unfortunate International Bitumen Company or Max Ball's equally troubled Abasand Oils Limited.

The original partners in Syncrude Canada Ltd. were Imperial Oil, the Canadian subsidiary of the Exxon Corporation of New York; Gulf Oil Canada Ltd., the Canadian subsidiary of Gulf Oil Corporation of Pittsburgh; and Atlantic Richfield Canada Ltd. and Canada-Cities Service Ltd., the Canadian affiliates of two small, but highly successful American corporations. Imperial Oil is the senior member of the Syncrude consortium. The largest petroleum company in Canada, it leases the greatest extent of acreage in the tar sands and, for that matter, in the Mackenzie River Delta. And by means of contracts given to another subsidiary, Esso Research and Engineering, Imperial has a large hand in the design, construction, and operation of Syncrude's refining plant. Syncrude was formed in 1965 to undertaken intensive research on oil-sands extraction. Four years later, when the Alberta government felt that synthetic crude production would complement, not compete with, conventional petroleum production, the consortium was granted a development permit by the provincial Oil and Gas Conservation Board to build an 80,000 b/d plant. In 1971, Syncrude applied for and received a 125,000 b/d permit, which required the consortium to give a firm decision to proceed with the project late in 1973. It duly did – but said, according to Alberta Premier Peter Lougheed, that the investment cost had gone from an early estimate of $600 million to, in the summer of 1973, $1 billion. Syncrude wanted to make a "go" decision, but was worried by the escalating cost and hesitant to make an outright commitment. To Syncrude, the whole project now seemed to be only "marginally" economic.

By the spring of 1973, Premier Lougheed of Alberta had become desperately anxious to see the development of the tar sands well under way and well ahead of any petroleum extracted from the oil shales of Colorado and also of cheap energy made available by nuclear research and engineering.

Otherwise, as he told the provincial legislature, the tar sands could be rendered "obsolete," even "useless." There was nobody willing to undertake tar-sands development except Syncrude, and something called an "energy crisis" was brewing. These are the main facts on public record, the rest being buried somewhere in government and corporate files. So all the reasons that caused the Alberta government to become an active participant in Syncrude are not known. But some conditions attached to this sudden public support of private enterprise are.

In the spring of 1973, the joint owners of Syncrude rejected an Alberta government royalty proposal based on the theoretical value of raw bitumen extracted from the tar sands. (In North America, oil royalties have been anywhere up to 25% of the wellhead value of a barrel of crude.) They argued that the GCOS operation had been crippled by the necessity of paying royalties during the years of development that produced no earnings. (This ignored the fact that GCOS had paid very low royalties to the Alberta government in certain years and no royalty at all in others.) Syncrude wanted – and somehow got – an arrangement whereby no payments would be made to the province until there were earnings, plus a guaranteed rate of return on investment. This latter would be calculated by deducting a certain percentage from the bulk of *total* investment – notably depreciation costs, interest payments, day-to-day operating costs, and depletion allowances – before arriving at yearly net profit figures to be shared 50-50 with the Alberta government. In effect, Albertans would subsidize part of Syncrude's investment risks, ostensibly via a 20% government ownership in the plant itself, this fifth, official owner being the Alberta Energy Company. (Such a profit-sharing scheme is a marvellous incentive to oilmen to get as much subsidization as possible by keeping tax-deductible costs up and taxable profits down.) The negotiations also included some remarkable "fringe benefits" for Imperial Oil and partners in the form of certain assurances by the provincial government. Changes in clean-air or clean-water legislation would be discussed with Syncrude before making these changes; licences to operate under these provincial acts would be given Syncrude not annually, but for five-year terms. Other extras were housing priorities in Fort McMurray and the paving of Highway 63 linking Fort McMurray and Edmonton. Imperial Oil and company also obtained a guarantee from the Lougheed government that there would be no strike or lock-out of labor while the plant was being built, even if legislation had to be passed to produce this convenient state of affairs.

All these concessions raise several interesting questions. Peter Lougheed, leader of the "Now" Conservative party of Alberta and premier of the province since 1971, is on record as saying that "Without a strong and vital oil-and-gas industry in Alberta, let's face it, we'd be a have-not province." Fair enough. It was American, not Canadian capital that built the post-Leduc oil-and-gas industry in Alberta. And it was to the midwestern United States that excess oil and gas was exported because, in the 1950s, Quebec and the Maritime provinces refused to accept Western crude that was more expensive than oil imported from Venezuela. (Ontario had to be persuaded by the federal government to accept the expensive Western brand.) In addition, Lougheed is determined to establish job-creating, wealth-producing secondary industry in his province, notably refineries and petrochemical products. Again, fair enough. In the years following Leduc, the Social Credit government of Alberta was content to let American oilmen siphon oil and gas out of the ground in astronomical quantities, pipe them south, and there process them into a thousand and one lucrative products. Nonetheless, in September 1973, why did Premier Lougheed appear on television to assure Albertans that the "long, hard, tough" negotiations with Syncrude had been settled "on the government's terms"? Did this camouflage a poor bargaining record during these negotiations? Was it an attempt to gloss over the awkward fact that certain costs had been passed by the private sector to the public sector? And where was any explanation of $600 million development costs becoming, by July 1973, $1 billion? All of which raises the most interesting, unanswered question of all. Who *really* owns the tar sands? Albertans or American oilmen?

Rather ironically, it was in Peter Lougheed's own province

that the enormous financial clout exercised by American petroleum corporations was closely investigated. About ten years earlier, Premier Ernest Manning and his Social Credit cabinet colleagues established a committee of enquiry under the chairmanship of Kenneth A. McKenzie, an Edmonton lawyer, to investigate and report on the marketing of gasoline. McKenzie and his colleagues began with an intensive examination of the petroleum industry and were simply staggered by the structure and scale of operation they found. In their opinion, very few members of the industry itself had the opportunity to see or comprehend "the whole elephant." In order to help outsiders begin to comprehend the petroleum colossus, the Commission used various comparisons. For instance, seven major oil organizations – Standard Oil Company of New Jersey (now known as Exxon), Royal Dutch/Shell Group (now Shell Oil), Mobil Oil Corporation, Texaco Incorporated, Gulf Oil Corporation, British Petroleum Company, and the Standard Oil Company of California – earned, as a group, more than the combined revenues of the governments of Britain, France, Germany, and Italy. In 1965, this close-knit business group (sometimes referred to today as the "Seven Sisters") owned, jointly and individually, 88.2% of all oil in the non-Communist world, whereas only 1.5% was state-owned. For almost four decades, these corporations had controlled the export of petroleum around the world and the marketing of refined products. In fact, members of this cartel allocated to each other this, that, and the next national market. The key to their set-up and success was their price for crude oil – an artificial price that enabled the cartel to extract fat profits from oil-hungry customers around the world.

> What was the market price of oil? The worldwide price of oil was the amount that was high enough to produce a profit on Texas crude, where production costs were then among the highest in the world. The price of all oil was a function of Texas prices. At the time the McKenzie Commission had the cartel under the microscope, it cost two dollars to produce a barrel of Texas crude. At the same time, it cost less than seven cents to produce a barrel of crude in Kuwait. Texas oil and Kuwait oil sold at the

same price on world markets! Since the greater part of the cartel-controlled empire, from the Middle East to Venezuela, was functioning on the peonlike labour standards of the underdeveloped world, the opportunities for profit-making were, of course, unparallelled. The cartel's own profit statements reflected the enormous disparity between American and colonial labour and production costs. In a year, for instance, when Jersey Standard earned six percent on its domestic assets, it made fifteen percent in Venezuela. While Gulf was earning ten percent on operations in the U.S., it was pocketing thirty-three percent from Kuwait. Just as the price of natural gas produced in Alberta was a function of what the American government wanted it to be, the price of crude oil all over the world was a function of what the American oil industry wanted it to be. The extortion has been going on for decades, first formalized by the industry's original big three – Jersey Standard, Shell and BP – in the Achnacarry Agreement of 1928. The big three were later joined by four other American cartel members – Standard of California, Texaco, Gulf and Mobil – in basing the prices of oil from all sources on the price of high-cost Texas crude plus transportation costs from Gulf of Mexico ports, and all seven were soon completely interlocked in joint production combines and joint marketing arrangements. . . .

Canada provided a copybook example of the artificiality of oil pricing under the pattern established by the cartel, which included the Canadian national oil policy [1961]. Under this policy, western Canadian oil is burned only by Canadians west of the Ottawa River, while their countrymen east of that line burn crude imported from Venezuela and the Middle East. Canada exports to the American West about ten barrels of oil for every seven imported into the Maritimes and Quebec.

At one point in the sixties, mathematicians of the Canadian Refinery Section of the Dominion Bureau of Statistics were startled to have to record that the oil that had left the wellhead in Kuwait at six cents per barrel and the oil from Alberta that had a production cost of $2.25 were reaching their respective markets in Montreal and Toronto at an identical price! Much of the oil used by Quebeckers comes from Creole Petroleum, Jersey Standard subsidiary and Venezuelan counterpart of Imperial Oil. This trade was at its height of profitability before the indigenous government was able to exert taxation and price pressures upon Creole, which currently pays a sixty percent royalty to the government. For years, the crude was leaving

Venezuela at a price close to $1.45 a barrel and arriving in Montreal at more than $3. It was the same oil, and the same barrel. But by the alchemy of mid-ocean transactions between Creole, an interlocked tanker company in Nassau and Imperial Oil, it was yielding the Jersey Standard empire more than 200 per cent in profit. Such artificiality reverberated to all Canadian consumers through the mechanism of the national oil policy. As the pressures of people in the underdeveloped world impelled cartel subsidiaries to hike their prices – Venezuelan prices have increased six times since the beginning of 1971 – the subsidiaries of the same cartel companies operating in Alberta hiked their prices to match.*

Between the fall of 1973 and the late winter of 1975, the Syncrude story becomes even more tangled. In the end, however, it wasn't just the citizens of Alberta who got taken to the cleaners.

Enter Herman Kahn, physicist, thinker extraordinaire, and a military strategist who has been described as "the Karl von Clausewitz of the nuclear age." A co-founder of one of the United States' most prestigious "think tanks," the Hudson Institute, Harmon-on-Hudson, N.Y., Mr. Kahn is not without impressive-sounding credentials. He is the author of books on how the U.S. can "win" a thermonuclear war: *On Thermonuclear War, Thinking About the Unthinkable*, and *On Escalation*. He has often dispensed personal advice on nuclear strategy to military organizations, including Canada's Department of Defence. In recent years, with the American absorption in such issues as domestic race riots and political extremism, plus a world fast running out of resources, Mr. Kahn has adapted to the roles of "social engineer" and "futurologist," witness his books *Things to Come* and *The Next 200 Years: A Scenario for America and the World*. Thus, on a continent becoming alarmed by increasing shortages – and the increasing expense – of petroleum supplies, it should not be surprising that he came up with one of his quickie solutions.

In November 1973, Herman Kahn travelled to Ottawa, where he delivered an economic sermon to Prime Minister

* Sykes, Philip, *Sellout: The Giveaway of Canada's Energy Resources*, Edmonton, Hurtig Publishers, 1972.

Pierre Elliott Trudeau, to Donald Macdonald, the Minister of Energy, Mines, and Resources, to a few senior civil servants from Macdonald's department, and to sundry other cabinet ministers. Exactly what Mr. Kahn preached in toto has never been substantiated, but the essence of his message has been verified. The "free world's" energy shortage could be solved by investing anywhere from $15 to $20 billion of international money in a crash program to construct twenty Syncrude-type strip-mining operations in the Athabasca tar sands by the late 1970s. These would produce a total of 2-3 billion barrels a day of synthetic crude, much of which would be exported for the first few years. Theoretically, none of this would cost Canada a penny: other national governments would put up all the money – provided Canada agreed to forego any substantial use of Athabasca crude for domestic purposes until the mid-1980s. As far as the engineering logistics of all this were concerned, Kahn theorized that the importation of 30-40,000 South Koreans would take care of the labor problem and that dumping about 10-20,000 tons per day of unprocessable, chemical-laden bitumenous wastes into the Athabasca River would get rid of the major disposal problem. (It would be a much better arrangement if Mrs. Kahn got a job and left her strange love at home to look after the house, where there would be no one to whom Herman could give advice about anything.)

Expose several thousand South Koreans and Heaven knows how many Albertans to culture shock? Kill the lower Athabasca River and contaminate the Mackenzie River, possibly all the way to Tuktoyaktuk on the Arctic Ocean? After listening to these remarkable ravings, Donald Macdonald, the minister of Energy, Mines, and Resources, was, as one newspaper reported, "underwhelmed." In June 1974, when Kahn gave a replay to the Alberta cabinet, the reaction of the provincial minister of the environment was that this was a scenario for social and economic chaos that would strain public facilities to such an extent and create so many chronic shortages of goods, material, and labor that there would be a "major citizens' backlash." However, Herman Kahn did not lack for converts and fellow-travelers.

Oilweek reported in its issue of December 24, 1973, that

Jean-Pierre Goyer, the federal minister of Supply and Services, stated "Government and industry should co-operate immediately to develop the Athabasca oil sands in time to sell to the U.S. while it is still oil hungry." Some weeks after the sermonizing in Ottawa, Goyer toured western Canada. On a couple of occasions he gave energy-policy speeches, parts of which sounded terribly like a blitz development of the Athabasca tar sands. That these particular speeches were made in the big-money centres of Calgary and Vancouver struck a few observers as significant, as did the fact that Goyer had set up Kahn's pilgrimage to Ottawa in the first place. Then, in the summer of 1974, the United States government – with which Kahn's think tank has close connections – made a curious suggestion to the Canadian government: the construction of a proposed Mackenzie Valley gas pipeline, a project favored by the Trudeau government, should be linked to a multi-billion acceleration of oil-sands development in order to arrive at mutually agreeable conditions of long-term U.S. imports of Canada's oil and gas. Also about this time in 1974, three petroleum groups approached the Alberta government for permission to build extraction and refining plants in Athabasca. Shell Canada Limited and Shell Explorer of Houston, Texas, planned a 100,000 b/d operation; Petrofina Canada Limited, Hudson's Bay Oil and Gas, Murphy Oil, and Candle Oil planned to work together on a 122,000 b/d operation; and Home Oil Limited – the only Canadian-owned company of the lot – and Alminex Limited, Toronto-based but Houston-controlled, proposed a 100,000 b/d operation. Each group projected production in the early 1980s.

In May, 1974, Ottawa had announced that royalties payable to provinces by petroleum companies would not, in future, be deductible from taxes due the federal government. This produced some curious reactions in the industry. Syncrude's anticipated petroleum production, which had been highly publicized as an export commodity, suddenly became vital to Canada's own domestic needs, particularly the need to become self-sufficient in energy supplies. At least, this was how Syncrude advised the National Energy Board. Shell Explorer, the U.S. affiliate that was partner to

Shell Canada Limited in a proposed tar-sands development, announced it was pulling out, partly because it could no longer be sure of exporting its share of bitumen-derived petroleum to American refineries. The Petrofina Canada-Hudson's Bay Oil and Gas-Murphy Oil-Candle Oil group talked of shelving their tar-sands plans, as did Home Oil-Aluminex Ltd. Several giant corporations - Imperial, Gulf, Chevron Standard - announced substantial reductions in explorations for gas and oil planned for 1975. And early in December, 1974, Atlantic-Richfield, one of the four Syncrude partners, withdrew from the project, despite the fact that several hundred workers had been operating on the site north of Fort McMurray for at least a year.

The decision of Atlantic Richfield to quit Syncrude hit Edmonton and Ottawa with all the force of a bomb. It shattered the Lougheed government's hopes of an ever-expanding economy and such future ventures as a planned petrochemical industry. It stunned the Trudeau government, already a bit groggy from National Energy Board (NEB) predictions a month or so earlier that Canada faced the threat of petroleum shortages by 1985. In the Board's opinion, Canadian oil could no longer be exported at the high rates that had prevailed since 1972-73, Arctic oil was not being discovered in significant quantities, and NEB's forecast of a continuing supply of Canadian oil after 1985 depended very heavily on the rapid exploitation of the tar sands. In fact, NEB's productibility graphs assumed that at least half a dozen Syncrude-type extraction plants would come "on stream" every second year between 1978 and 1993.

A second bomb went off a few weeks later at a press conference held in Toronto. Imperial Oil, Gulf Oil Canada Ltd., and Canada-Cities Service Ltd., the three remaining Syncrude partners, speculated in the course of a press conference that, to keep the project alive, either more money had to come from somewhere - quite possibly from the public purse - or there would have to be special tax concessions. Since the project was already expected to cost $1 billion, how much was "more money"? "We'd like to get another billion," replied one company president. These details emerged in the course of the question-and-answer session with the media in Toronto on January 16, 1975. And, the consortium said, the deadline for acceptance or rejection was January 31. If accord was not reached by then, the project would have to be closed down.

As it finally turned out, that wasn't all the three companies wanted. As during the 1973 Syncrude-Lougheed negotiations, a guarantee would have to be obtained from the federal government that synthetic crude from the tar sands could be sold at the world oil price, not at the lower Canadian one. Syncrude's owners were determined to sell an Alberta-owned resource at energy-crisis prices! And, also as demanded by Syncrude in 1973, there would have to be a guarantee of the exemption of synthetic crude from the recent federal taxation policy (due for implementation in 1976) of non-deductibility of oil profits paid to provincial treasuries.

When Finance Minister John Turner's budget in November, 1974, confirmed that oil royalties paid to provinces would not be deductible, newspapers had been quick to print scare stories. Oil-field equipment and supplies in Alberta were being hurriedly transported south to the United States in the dead of night; people in Edmonton and Calgary suburbs were discovering that oil-industry neighbors had been transferred to Dallas, Texas, or some other such petroleum centre. This time, however, the press blew its collective cork and went to bat for Canada. In Toronto's *Globe and Mail*, a columnist angrily asked, "Can any self-respecting government permit itself to be rushed and forced into throwing a huge amount of taxpayers' money on the table just because the other players, a trio of oil companies, are threatening to take their cards and chips and go home? Somehow that smacks more of blackmail than poker." The Calgary *Herald* was moved to editorialize " . . . the long and short of it is that the Canadian government has been told to change gears by three American oil companies. That simply does not sit well even with those such as the *Herald* who have argued long and hard against many of Ottawa's energy policies." There was a great deal of editorial support for some sort of public ownership of the tar sands. In fact, Bruce Willson, a long-time oil-and-gas man, was so appalled by

Syncrude's terms that he publicly demanded outright nationalization of their project. However, maybe the shrewdest comment of all was made by another member of the oil industry. He didn't believe for one moment that the Syncrude group really wanted more money and a couple of guarantees. He could envisage the long-term benefits of a venture that included the dollars of government, provincial and federal. If there is a considerable amount of public money invested in a commercial project, do such things as product price, distribution, and sale, plus, for instance, environmental disruption or chaos, really matter? Aren't the governments in question obliged to report success, whatever the ultimate cost, monetary and otherwise? Further, suppose that those observers who say there is a *surplus* of petroleum in the world are right. Suppose, in time, the price of petroleum goes *down*? Isn't it smart to get the public purse to pick up some of that particular loss? Alberta oilman Nick Taylor probably hit the nail right on the head when he compared arranging an oil venture with government to "marrying the landlord's daughter. You stop paying rent, get the first pork chop, and the best bed in the house."*

So what happened on January 31st, 1975? Nothing. It all happened in the course of the weekend of February 1-3rd. In a hotel room in Winnipeg, Manitoba, the governments of Canada, Alberta, and an energy-hungry Ontario said "Yes." Their representatives accepted Syncrude development cost estimates that had somehow escalated from $1 billion in mid 1973 to just over $2 billion by the end of 1974. (Inflation is tough on all of us, but it isn't that tough. Obviously, if you increase front-end costs – a portion of which Syncrude is allowed to deduct from net profit – this reduces the amount of net profit you share on bitumenous crude.) Despite a budget that forbade the deduction from federal taxes of payments to provinces for the extraction of oil and natural gas, Messrs. Trudeau and Turner promised Syncrude exception from this ruling. Indeed, the Liberal cabinet proved itself an ultra-liberal group of thinkers. It even promised that the project's synthetic crude could be marketed at world

* Pratt, Larry, *The Tar Sands: Syncrude and the Politics of Oil*, Edmonton, Hurtig Publishers, 1976.

prices, give or take certain adjustments to do with transportation and grades of crude produced.*

And the extra money? Sure enough, some of it comes from the public treasury. The three companies are committed to spend more of their own money to achieve the $2 billion expenditure. However, Ottawa joined for 15% interest, which amounts to paying out $300 million; and Ontario joined for 5%, $100 million. In addition to providing 10% of the risk capital, Alberta agreed to pay for certain plant costs to the tune of an estimated $300 million and also to bear 80% of the cost of building a pipeline to Edmonton. In return, three governments share a 30% interest that will pay no return on their joint investment until such time as millions of oil-company dollars in depreciation costs, interest payments, operating costs, depletion allowances (and the fixed-rate-of-return-on-investment guarantee) have been taken into account. And, of course, these companies can reduce the true cost of their 70% investment via Syncrude's tax classification as a mining, not a conventional oil-well operation.

One unfortunate aspect of the Winnipeg deal is that it sets a precedent. As Canada's self-sufficiency in petroleum energy weakens and we have to import more and more oil from abroad, what's to stop Shell Canada from demanding – and getting – as sweet a tar-sands deal as Syncrude? What's to stop Home Oil-Aluminex doing the same thing? If there is ever a huge oil or gas find in the Mackenzie Delta or on the Arctic islands – all of which are thousands of miles beyond Athabasca – why shouldn't the American-controlled petroleum companies exploring there demand a costly, transportation subsidy à la Syncrude to get this oil or gas to market?

There is, however, a positively terrifying aspect to the Syncrude project that has nothing at all to do with dollars. If Herman Kahn did nothing else, at least he raised this spectre with his glib talk of a massive industrialization of the Athabasca tar sands.

---

* In late June, 1977, the Trudeau government publicly confirmed that Syncrude Canada Ltd. would obtain world prices for its synthetic crude production via a subsidy of $125 million handed out during the period 1978-1980.

The National Energy Board wasn't the only Canadian authority to assume the exploitation of the sands in order to maintain national self-sufficiency in petroleum. In the early 1970s, the federal Department of Energy, Mines, and Resources also looked into the future and came up with its predictions of future supplies of oil. These postulated significant discoveries of petroleum both off the Atlantic coast and in the Arctic to counterbalance declining production from conventional wells in western Canada. But the really interesting thing about the Department's estimates was their basic assumption that today's methods of strip-mining the tar sands could provide 2 million barrels of petroleum a day by 1990. Indeed the Department figured that, by the year 2000, this production could be doubled (a great comfort to various governments in Canada, which have long been committed to earning dollars by exporting petroleum every day of the week to the United States). These calculations, theoretical though they are, seem plausible because there *is* a technology to mine the tar sands, and synthetic crude *can* be easily pumped to Edmonton to link up with several transcontinental pipelines. The snag is that, in order to boost tar-sands production to 2 million b/d by 1990, there would have to be sixteen Syncrude-sized (125,000 b/d) plants. To boost it to 4 million b/d, there would have to be an additional sixteen. Shades of Herman Kahn! One plant built every year between 1974 and 1990 and one every six months in the next ten years! Such a gargantuan development is out of the question. Where would the capital, labor, and materials come from for thirty-two Syncrudes? Where would Alberta find the money to build, in effect, a mini-province within a province: a massive network of roads, highways, hospitals, police and fire protection, schools, housing, and so on?

But consider for a moment a somewhat less ambitious plan to complete one Syncrude-sized plant every four years for a total output of 1 million b/d by the year 2000. This plan was put forward in 1972 in an Alberta civil-service report as a realistic crash program of development. (With one extraction plant already at work, one "coming on stream" in 1978, and three others waiting in the wings, so to speak, by 1990 there could well be five tar-sands operations going full blast in Athabasca.) The social and economic costs of eight Syncrude-type operations will be tough, perhaps even impossible to achieve. However, the environmental costs would be even worse.

On average, somewhere between three and five tons of bitumenous sands have to be moved and processed for each barrel of synthetic oil produced. So about half a million tons of earth, stone, and sand would have to be dug up each day to support each of the eight plants. Vast though this chore is, it's not the problem. The trouble is that millions of cubic feet *a day* of tailings – contaminated water from the hot-water extraction process – would have to be stored in settling ponds. This water contains, among other things, fine clay particles, some oil, plus certain phenols, chlorides, nitrates, and sulphates. A small portion of the water can be recyled for use in the extraction process, but most of this poisonous mixture has to stay sitting in these ponds for long periods to allow all its contents to settle. Nobody has yet found any other way of disposing of it. This polluted water is bound to seep into underground water tables, yet nobody knows what long-term effects this will have on the already sodden environment of a sub-arctic region. (The GCOS tailing ponds are right next the Athabasca River!) Air pollution accompanies tar-sands exploitation. Bitumenous-derived petroleum has a high sulphur content, which has to be removed in the plant refinery by a coking process that releases enormous amounts of sulphur dioxide (SO_2) straight into the atmosphere. Eight plants would emit several thousand *tons* of SO_2 a *day* into the air and create an infinitely worse version of Sudbury, Ontario, where the processing plant of the International Nickel Company of Canada has wiped out all vegetation and produced a lunar-like landscape. (So lunar-like, that U.S. Moon-exploration teams did part of their training at Sudbury.) Bad though this is, there is another, equally fearsome fallout. The Fort McMurray region is subject to numerous temperature inversions, a weather condition in which a layer of slow-moving, almost stationary, warm air overlies a layer of colder air and literally imprisons that air – and any pollutants in it – close to the ground. And SO_2 combines with the water vapor in the air to form sulphuric acid, a highly corrosive substance that was a major constituent of "killer fogs" that plagued such places as London and Tokyo.

The Spanish have a shrewd saying: "Take what you want from life, but be prepared to pay for it." Perhaps the real cost of intensive tar-sands development will not be just a further stage in Alberta's conversion into what one writer has called "a veritable New Jersey of the north crisscrossed with energy corridors and roads to non-renewable resources."* Perhaps the real payment will be the conversion of the lower Athabasca River and the nearby Lake Athabasca – the "top" of the Mackenzie's "trunk" – into a biologically barren wasteland.

* * *

Oil and natural gas do a tremendous amount to make life cosy and comfortable for Canadians and Americans. They heat our homes, offices, and factories. They power hundreds of different kinds of machinery, the most important of which to many of us is the automobile. They are the basic "feedstock" from which the petrochemical industry manufactures thousands of synthetic products ranging from aspirin to waxes. As one journalist sums it up,

> If you were to calculate the total amount of energy produced by various kinds of machines in Canada and the United States and convert it into human muscular power, you would find that every Canadian and American has the equivalent of about five hundred human "slaves" working for him. All that energy – all that ability to do work – has to come from somewhere: if it doesn't come from real slaves, it has to come from raw materials like coal and petroleum.

But the same writer goes on to note something else.

> Slowly but surely we are beginning to run out of many of these resources which, once used, cannot be replaced. Resource consumption, and therefore waste production and pollution, increase exponentially. That is, they tend to double and

---

* *Ibid.*

> redouble at regular intervals over time – intervals that can be predicted if you know the rate of growth. If you plot an exponential curve, you find that for a long time it grows deceptively slowly. As the base of each new doubling gets larger, the curve gets steeper. And then it suddenly shoots upward right off the graph. All the information I had access to indicated that in a frightening number of cases of resource depletion and pollution levels, we were at that point on the curve where it begins suddenly to go vertical.*

Throughout the "petroleum age" in human history, there have been predictions that supplies of oil and natural gas were in danger of imminent exhaustion. Are today's forecasts of shortages in the same gloom-and-doom category? Or are they really getting at the truth?

In the 1940s, forecasters of shortages totalled up existing North American reserves and subtracted the then-current rate of consumption to arrive at what they called a "reserve life index." They made no allowance for year-by-year fluctuations in the reserve life index when new reserves, greater than a year's usage, were discovered. Thus, in the years following 1945, the reserve life index was either increasing or remaining steady. For every billion barrels of oil or trillion cubic feet of natural gas used up, at least a billion more barrels or a trillion more cubic feet of new reserves were being found somewhere on this continent. In recent times, however, the reserve life index has been steadily declining: demand has been much higher than the rate of discovery. In the 1960s, the U.S. was forced to import more and more oil and natural gas, mainly from the Middle East but also from Canada. In 1970, Alberta's rate of production of natural gas equalled the rate of discovery. In 1974, the Canadian Petroleum Association noted in an April issue of the *Oil and Gas Journal* that "Canada's liquid hydrocarbon reserves took their biggest drop ever in 1973, declining for the fourth consecutive year. . . ." By the same date, domestic reserves of U.S. oil were in their fourteenth successive year of decline. Why? A major reason is that, exponentially speaking, even a 4% growth in consumption doubles consumption in 18 years.

* Rowland, Wade, *Fuelling Canada's Future*, Toronto, Macmillan of Canada, 1974.

The current energy crisis in North America is not simply the result of more people using more energy. Consumption of oil and natural gas continues to grow at about 4% a year, but this growth had been foreseen. The energy crisis also stems from unforeseen changes in the *supply* of fossil-fuel energy. In North America, decades of low prices for oil and gas discouraged searches for new resources while steadily depleting existing supplies. However, in the Middle East and South America the days of dirt-cheap oil are gone. No longer can American-dominated petroleum companies take oil out of Saudi Arabia or Libya or Venezuela at the cost of a few cents a barrel and sell it around the world for a couple of dollars a barrel. The member countries of the Organization of Petroleum Exporting Countries (OPEC) have been able to secure higher taxes and royalties on what is, after all, their petroleum and obtain larger shares in the ownership of pipelines and petroleum reserves on their lands. But if petroleum is to continue to carry the heaviest burden of energy demands, then it has to be sought in "frontier" regions. Oil companies such as Exxon, Gulf, Texaco, Mobil, Shell, and British Petroleum need giant new fields to keep them in business in the 1980s and 1990s. It is for these reasons that drill rigs dot the wind-whipped surface of Europe's North Sea, loom out of the fog-bound waters of "Iceberg Alley" off Labrador, interrupt the snowy monotony of the Arctic landscape, and rear above the leaden seas off the western coast of Nova Scotia.

It was in 1968 that one of the giant new fields so desperately needed by oilmen was located. A tremendous flow of natural gas erupted from a wildcat well being drilled on the narrow, Arctic-coast plain of Alaska by the Atlantic-Richfield Company of Philadelphia and by Humble Oil and Refining Company, a subsidiary of the Exxon Corporation. Additional wells were drilled to delineate the extent of the field, and Prudhoe Bay gave every indication of being the largest oil field ever discovered in North America – roughly about 10 billion barrels of recoverable oil reserves, plus tremendous amounts of natural gas.

The Prudhoe Bay discovery started an invasion of the nearby Mackenzie River delta that was as feverish and frantic as any gold rush. By 1970, firms working in the Delta included the French-controlled Aquitaine and Elf, Numac Oil and Gas, Gulf Canada, Shell Canada, and Bow Valley. Within twelve months of the Prudhoe find, the amount of land in the Canadian north held by petroleum companies under exploratory permit from Ottawa rose from 180 to 320 million acres. By mid 1973, the figure had soared to 844 million. Imperial Oil, which leased about 10 million acres under permit from Ottawa in the Mackenzie Delta, announced in 1970 a considerable discovery of oil at Atkinson Point. Imperial immediately began its "step-out" work, drilling another well nearby to help establish the scope of the find, and also began exploring elsewhere in the Delta.

Gas finds, however, occurred more rapidly than oil discoveries. As early as 1972, the petroleum industry was making statements to the effect that the total of reserves required to support a gas pipeline from the Delta all the way to markets in southern Canada and the northern United States had been found – or soon would be. By this time, the American government had rebuffed a Canadian government suggestion that Prudhoe Bay oil be piped out to markets in the U.S. Midwest via the "corridor" of the Mackenzie Valley. Washington announced the construction of an all-Alaska oil pipeline route to Pacific tidewater at the deep-water port of Valdez. Still, American petroleum companies were most interested in Mackenzie Delta gas and lavished millions of dollars on pipeline-design studies and test stations in the Delta. Exploration even moved beyond the Delta onto the inshore waters of the Beaufort Sea, where artificially-created islands enabled drilling crews to probe beneath the sea floor. As one oilman put it, "That's a big piggybank up there. We've put hundreds of millions of dollars in and haven't taken a thing out yet."

Ottawa, too, responded to the post-Prudhoe Bay activity in the Delta. The Trudeau government produced a document entitled "Preliminary Pipeline Guidelines for Oil and Gas in the Mackenzie Valley," which indicated considerable cabinet interest in a Mackenzie corridor for oil and gas, Alaskan and Canadian. These guidelines also strongly hinted that only *one* pipeline of the three then being proposed would receive official approval. The Department of Indian Affairs and Northern Development (DIAND), which has an almost province-like authority to administer, regulate, and control the lives of Canadians living north of 60° - which is anywhere in the 1.3 million-mile mass of Canada that sprawls north of the ten provinces - established a Commission of Inquiry into the terms and conditions under which construction of a natural gas pipeline would be allowed in the Mackenzie Valley. The Commission's chairman, Mr. Justice Thomas R. Berger of the Supreme Court of British Columbia, spent almost three years acting as a one-man tribunal investigating the pipeline that had been much-publicized as the answer to Canadian - and American - energy needs as far forward in time as the year 2000. According to DIAND, this was not an investigation to determine if a pipeline should be built but under what conditions it *would be* built. However, Mr. Justice Berger gently but firmly rapped Ottawa over the knuckles on that one. He refused even to begin the investigation until officialdom, including the Trudeau cabinet, agreed that his brief was to examine the whole way of life in the North - pipeline or no pipeline.

Provincial governments got into the fossil-fuel business in the Delta. Alberta and British Columbia, anxious to expand their industrial bases by developing petrochemical industries, invested in the Alberta Gas Trunkline Company Limited and West Coast Transmission respectively. These companies, together with other corporate groups, were backers of the "Maple Leaf" or "all-Canadian" pipeline, a proposal by Foothills Pipe Lines Limited to build a small-diameter (36-inch) gas pipeline from the Delta to southern Canada. The opposition to this was Canadian Arctic Gas Pipeline Limited, a U.S.-dominated group that proposed to build a large-diameter (48-inch) pipeline carrying both Alaskan and Delta gas to Canadian and American markets.* Ontario formed an Ontario Energy Corporation and considered investing in Mackenzie Delta-Beaufort Sea exploration programs.

* Participants in this project, in 1977, are listed on p. 178.

Nobody seems to know just how much gas and oil is recoverable from the Mackenzie Delta and from beneath the floor of the Beaufort Sea. One set of impressive estimates is available from the Canadian Petroleum Institute. The Geological Survey of Canada has another, much smaller set. The National Energy Board has been steadily revising its totals – downwards. Various oil companies have also revised theirs – upwards. None of which is as confusing as it sounds. Enough significant finds of natural gas have been made near the mouth of the Mackenzie River and below its offshore waters to make it reasonable to assume the ultimate discovery of other considerable reserves. Every known ear-mark of a major petroleum basin has been found in the Delta from surface seepages of oil and bitumen to deep-drilling samples of reservoir rocks of sandstone and limestone origin capable of containing billion-barrel deposits of petroleum. The real question is: How does Mackenzie/Beaufort Sea gas get to markets thousands of miles away? The physical, technical, and financial problems involved in answering that question make building the

---

Canadian Arctic Gas Pipeline Limited, Toronto.

Majority-owned Canadian companies:
Alberta Natural Gas Company Limited
The Consumers' Gas Company
Northern and Central Gas Corporation Limited
TransCanada PipeLines Limited
Union Gas Limited.

Minority-owned Canadian companies:
Gulf Oil Canada Limited
Imperial Oil Limited
Shell Canada Limited.

American-owned companies:
The Columbia Gas Transmission Corporation
Michigan Wisconsin Pipe Line Company
Natural Gas Pipeline Company of America
Northern Natural Gas Company
Pacific Lighting Gas Development Company
Panhandle Eastern Pipe Line Company
Texas Eastern Transmission Corporation.

As of Oct. 31, 1975, the Canada Development Corporation (CDC) withdrew as a paying participant and continued its affiliation as an associate member. The CDC's conditional commitment to invest $100 million in Arctic Gas equity-related securities remained in effect.

Canadian Pacific Railway or constructing the St. Lawrence Seaway seem like minor-league enterprises.

When this fuel is piped down part or all of the corridor of the Mackenzie River, it will be brought through a singularly unusual part of Canada: a desert. It receives very little rain and snow per year. On the basis that ten inches of snow equals one of rain, total annual precipitation is just under ten inches for most of the northern mainland of Canada. It is also a frozen desert, above ground in winter and below ground winter and summer. The popular image of Canada as a land of ice and snow has a solid basis in fact. Permafrost, permanently frozen ground, underlies most of the 40% of Canada where the provinces end. At the surface, a blanket of soil anywhere from two to ten feet deep freezes and thaws with the seasons, but below it is iron-hard ground. This legacy of the Pleistocene era often takes the form of solid ice. In well-drained layers of sand or gravel, it is a sort of dry deep freeze. Dig a pit in permafrost soil, cover the opening with an insulating layer of peat or mosses, and you've got an instant, year-round refrigerator. (In the early days of mineral extraction in Mackenzie River territory, miners had to thaw uranium- or gold-bearing ore before they could get it out of the ground.) Detailed knowledge of permafrost distribution and depth in Canada is lacking. However, what little study has been done suggests fairly consistent thicknesses of at least 1,000 feet over the northernmost half of the Northwest Territories and the entire Mackenzie Delta, and large and small patches about 200 feet in depth in the Mackenzie Valley and in most of the Yukon Territory.

Permafrost is a natural force all on its own. Its built-in pressures nudge soil materials to the surface to form blister-like mudspots and protusions of silt. Sometimes there are huge eruptions below ground that throw up peculiar structures called pingos, ice-mounds that look like monstrous boils on the face of the land. Even in its normally inert state, permafrost can produce startling effects. The long hours of northern sunlight warm the surface of the ground, and exposed soil thaws into something that has the consistency of a sopping-wet sponge. On the gentlest of slopes, a section of earth will loosen itself and creep downhill over a slick, slippery foundation of ice. Trees and shrubs, whose roots cannot penetrate ice-encased soil and anchor them, tilt drunkenly or fall with the wind.

Man's failures to come to terms with permafrost have been legion. Any building not elevated above the surface of the ground on wooden or concrete piles (or on a gravel foundation several feet thick) causes meltings and refreezings that change the contour of the ground. Structures are either pushed aside or their floors and walls twisted and warped. Remove forest and bush cover to make a road and, without a gravel base, you end up with a trench full of mucky water. Treadmarks on permafrost have been known to expand into mud sloughs that grow larger and larger each year due to the effects of radiant heat from the sun and the erosive nature of the surface collections of water. (Bulldozer tracks made in the north twenty or thirty years ago when radar lines were being built are still clearly visible today, scars that criss-cross the face of the land.) Once an area of permafrost thaws because of the presence of heat and/or the removal of the shallow, insulating layer of surface vegetation, an endless process of erosion begins – and spreads in all directions. Erosion even seems to ripple through subsurface layers in patterns still not very well understood.

The technical and financial difficulties involved in building a 2400-mile, $7-10 billion Mackenzie River natural gas pipeline are well-illustrated by the construction of the Prudhoe Bay-Valdez oil pipeline in neighboring Alaska. Only 800 miles in length, it's a real horror story.

In 1923, a section of northwestern Alaska almost the size of the state of Virginia was set aside by the U.S. Navy as Naval Petroleum Reserve No. 4. Here, in the shadow of the icy palisades of the Brooks Range, a few holes were punched through the tundra surface of "Pet 4" in the 1940s when oil shortages threatened to hinder the Allied war effort. Despite Eskimo reports of mysterious black lakes that burned, no drilling of any significance took place on the North Slope of Alaska until the winter of 1967-1968. A few miles east of "Pet 4" at Prudhoe Bay, where the Sagavanirktok River empties into the Beaufort Sea, Atlantic Richfield and Humble Oil found rich petroleum deposits 10,000 feet below the tundra. A year later a neighboring company, British Petroleum, announced a similar find. Something like 10 billion barrels of oil and 27 trillion cubic feet of natural gas had been tapped. Almost immediately, predictions of yet-to-be-found offshore deposits in the Beaufort and Chukchi seas hit astronomical highs.

In 1973, the Alyeska Pipeline Service Company, a consortium of eight petroleum corporations, was finally allowed to begin constructing an 800-mile pipeline across three mountain ranges and under some 350 streams and rivers to channel 600,000 barrels of oil a day from Prudhoe Bay to Valdez for shipment to refineries in the states of Washington and California. Years of political and legal battles among natives' rights, conservationists', citizens' groups, and oil company lobbyists in Alaska and in Washington, D.C., had complicated the rules and, inevitably, the price of construction. Inflation, plus labor-union demands, also caused prices to jump. (The line was completed and became operational in the summer of 1977. The cost was almost $8 billion, making the project the most expensive construction job in human history.) But it was another, quite unanticipated, controversy that helped to quadruple initial costs: the builders' haste to complete the pipeline project resulted in extensive violations of strict environmental and technical regulations.

The combination of the rugged face of Alaska and the many environmental and technical regulations stipulated in right-of-way agreements signed by the Alyeska Pipeline Service Company, the State of Alaska, and the U.S. Department of the Interior resulted in the most sophisticated pipeline ever designed for use anywhere in the world. For almost half its length, the heavily-insulated, 48-inch line is laid on top of posts planted 25-feet deep in permafrost so delicate that a one-degree increase in temperature can change it into a brackish muck. (These posts are filled with liquid ammonia to keep the surrounding soil frozen.) Pumping stations at fixed intervals pressurize the oil to keep it free-flowing at temperatures around 130°F while surrounding air temperatures range from 90°F to -60°. Every 1,000 or so feet, teflon-coated "shoes" allow the pipe to slide sideways on special support beams to compensate for thermal expansion and contraction and also for earthquake vibrations. At these support points, fibreglass sheathing and polyurethane panels insulate the pipe to keep the oil heated in the event of a pumping-station breakdown. Sections of pipe that would block caribou migration routes are buried and the ground surrounding them kept rigid by a combination of extra polyethylene insulation and refrigerated brine pumped through small pipes around the main line. Where the pipeline plunges under streams and rivers, it is buried five feet below the bed of the waterway and then "jacketed" in concrete. Remote-control valves spotted along the line can close off the flow of oil within a section of pipe in the event of a break (although, in the four minutes it takes to shut down the flow, a spill of several thousand barrels would occur).

By mid-1974, several task forces of men and women had begun building an all-weather haulage road across Alaska and laying the 50-foot-wide gravel work pad to protect the land during the actual construction of the pipeline. Alyeska controlled the budget and the engineering planning side of things, while the giant Bechtel Corporation, working through hundreds of subcontractors, was responsible for production and quality control (and got its usual no-strike agreement from state and union officials). A year later, the installation of the pipeline was under way, thousands of workers laboring twelve hours a day, seven days a week in order to complete half the line by December, 1975. Trouble was inevitable, and manifested itself in various forms. It varied from inter-union violence to the abuse and theft of machinery and supplies. Time and again, state and federal officials reported blatant violations of stipulations safeguarding waterways and wildlife. However, worst of all was the revelation in 1976 of faulty welding of the 80-foot sections of pipe: X rays of pipe welds confirmed that 10% of some 40,000 welds required repair or replacement. Alyeska reluctantly admitted that production co-ordinators had been over-riding the authority of quality-control inspectors: tight construction schedules had priority over environmental promises. Worse still, many X rays taken by Alyeska had been tampered with or falsified to a point where it was impossible for anyone to certify whether or no *all* 1975 welds had been examined and accounted for and *all* defects noted. Utimately, after the U.S. Congress had been forced to send a team of investigators to Alaska, hundreds of workers and machines began excavating huge holes in the iron-hard ground so that questionable welds could be X-rayed and certified as safe, or re-welded. North of the Brooks Range alone, this involved digging up 1,000 weld points. To complete everyone's misery, at the Sagavanirktok River, several hundred feet of pipe rose to the surface, its concrete straitjacket fragmented by intense ice and permafrost pressures.

There are two peculiar aspects to the Alyeska Pipeline, the second of which has particular significance for Canadians. First, the Prudhoe Bay oil that is being pumped through the line won't *begin* to match the 60% increase in U.S. crude oil imports recorded between 1973 and 1976: Americans are currently using 18 million barrels of oil daily, so how long are 10 billion barrels going to last – and at what price per barrel? (The Prudhoe Bay-Alyeska oil companies aren't going to absorb the $2,000,000-a-day interest charges on pipeline loans. Remember how Syncrude persuaded Canadian governments to pick up 30% of the tab for its tar-sands operation? The powerful U.S. oil lobby in Washington will find a way to stick these charges to American customers.) Second, Prudhoe Bay natural gas – trillions of cubic feet of the stuff – is still up there at Prudhoe Bay, and a line has to be built to get it to southern markets. There has long been talk in the oil industry of building another pipeline to Valdez, or piping gas in a line alongside the Alaska Highway, or routing it via the edge of the Beaufort Sea (or the Yukon interior) to the Mackenzie Delta. Which is where a Mackenzie pipeline really comes into sharp focus.

To justify even a small-diameter pipeline from the Delta, industry experts agree that 15-20 trillion cubic feet of gas must be available for piping. The highest estimates of Delta gas are in the 5-7 trillion-cubic-feet range. Prudhoe Bay reserves are much larger (27 trillion) but are mixed in with oil. Under Alaska conservation law, natural gas can only be taken out of the ground as the oil is extracted. Thus, gas extraction will only build up in volume as does the flow of oil. It will take only a year or two, maybe three, for Delta gas to go through a Mackenzie pipeline. After that, there's an awful lot of Prudhoe gas available for pumping right through the same line, into connecting lines – and onward into the United States.

* * *

The Alyeska experience will be repeated in the Northwest Territories and/or the Yukon Territory. There's an inevitability to this that is implicit in the gullibility displayed by Ottawa politicians and civil servants, together with the influence exercised by petroleum companies. These two groups have confused and conned Canadians into believing that black is white and white is black. An extreme statement? Not when you recall the following facts.

In 1969, the National Energy Board issued a statement based on industry-supplied figures of petroleum reserves. NEB assured Canadians that, by 1985, the nation would be exporting at least 900 million barrels of oil each year to the United States and, by 1990, one and a half billion barrels annually. Remember "Honest" Joe Greene, the federal minister of Energy, Mines, and Resources? In June, 1971, using statistics that originated partly with the Canadian Petroleum Association, Mr. Greene gave a speech to the Canadian Institute of Mining and Metallurgy in which he stated that, at 1970 rates of production, Canada had "923 years of natural gas and 392 years of oil" in the ground. Not surprisingly, by 1973, Canada was exporting about 58% of its oil production and 45% of its gas production to the United States. Everything looked great. (Nobody publicized the fact that these enormous exports were taking place while U.S. oil-and-gas reserves were rapidly shrinking and the OPEC countries were strongly objecting to the dirt-cheap prices given them for their oil.) Our happiness was confirmed by the federal Department of Energy, Mines, and Resources, which published in June of that year *An Energy Policy For Canada:* its message was that Canada could meet its oil needs until at least 2050 A.D. – and at prices less than $7.00 a barrel. What Canadians did not know until a year later, via an *Oilweek* interview with the then Energy Minister, Donald Macdonald, was that their government was, in his words, "virtually dependent on the major [petroleum] companies for the sources of information."

The crunch came that same year. The Yom Kippur War of June 1973 taught certain Arab states the political uses of petroleum: they cut back oil production to put pressure on supporters of Israel around the world, notably the United States. By December, 1973, the Organization of Petroleum Exporting Countries – which is mainly Arab in membership – had upped the world price of oil from $3.00 to $11.00 a barrel.

Suddenly Calgary and Toronto oil spokesmen were saying that if the price of domestic oil did not begin to rise to the world price, Canadians would face oil shortages by 1980. If their companies were not given "adequate incentives" to explore for new reserves of oil and gas, we'd all freeze to death. In the dark! They were willing to look for new oil reserves in the north, but they weren't going to spend any money on exploration and development without these "adequate incentives," a euphemism for higher prices. (And this despite the fact that, under Canadian tax law, oil companies can write off 100% of exploration expenses against corporation tax.) So they pressured the Trudeau government to raise the price of domestic oil. Do you remember what happened shortly thereafter? By the spring of 1974, the Canadian price of $3.80 a barrel had shot up to $6.50 a barrel. Of course, some of this increase (and of subsequent price hikes) was used by Ottawa to help subsidize the Atlantic provinces, who were now paying a steeper price for Venezuelan oil. But what oilman bothered to suggest an extension of the Alberta-Toronto oil pipeline to Montreal, if not into the Maritime provinces? In October of that year, NEB announced – again with some help from oil-industry statisticians – that there would be no supply of domestic oil to outlets west of the Ottawa Valley after 1982. By July, 1975, the price of domestic well-head crude was $8.00 a barrel. (In the period 1972-1975, natural gas went from $0.18 to $0.90 per thousand cubic feet.) In 1975, Canada began importing more oil than it was exporting. In April 1976, Alastair Gillespie, Minister of Energy, Mines, and Resources, issued a report to the effect that Canadians had 60% less petroleum than they were said to have had in the same Department's report of 1973 and he predicted purchases of foreign oil running into billions of dollars. Do you recall what happened shortly thereafter? In July that same year, the Trudeau government upped the price of oil to $9.05 a barrel, partly in order to help the oil companies help Canada become two-thirds self-sufficient in oil production by the mid-1980s. By January, 1977, the price was $9.75 – a 275% increase in less than four years. And in June, 1977, Alastair Gillespie announced a further series of price hikes that will increase the cost of a barrel of oil another 41% by 1979.

All of which makes you wonder what happened to Joe Greene's 923 years of oil and his 392 years of gas, which industry statisticians and government spokesmen say have shrunk to about 7 and 20 years respectively. Imperial Oil commented on the discrepancy in figures in a 1976 press release that quoted Executive Vice-President D. K. McIvor on the Canadian Petroleum Association's (CPA) estimates of reserves in the late 1960s.

> With the benefit of 20/20 hindsight, I can see where I might have disputed the CPA estimates. But because of the relative lack of hard exploration and development data from the frontiers, all the estimates made at that time were speculative and it was virtually impossible to support or deny them on the basis of hard fact.
>
> While we thought that we had a more sophisticated assessment technique, assessing reserves is not an exact science, and there was always the possibility that the CPA figures were right and we were wrong. Publicizing our numbers might only have confused the issue.
>
> Besides, even if I had had the foresight to express discomfort, for competitive reasons I would probably not have set out our exact numbers. By the late sixties we had begun to get very active in the frontiers, in competition with other members of the industry. We thought we had more precise numbers, and we did not want to have to start bidding against our own data.
>
> In the minds of some people this may seem like more evidence against the competitive system. In fact it is competition that has stimulated the search for Canada's oil and gas reserves.
>
> Oil explorers are gamblers. Like all gamblers, the individual explorer knows the odds are in favour of the casino. At the same

time he believes that he has a better system and, while the other players may lose, he is going to win.

In Canada there are several hundred companies engaged in exploration – all with different perceptions at various times. Company A, on the basis of its information, may walk away from a prospect. Company B, with different information, may take a look at the same prospect, decide that it likes it, drill, and find oil or gas. Without competition these discoveries might not be made.

We are always going to have critics who insist on seeing something sinister in our actions – and who will insist on believing that we remained silent in a deliberate attempt to rig reserve estimates in order to encourage exports. I suggest that the behaviour of the industry simply does not square with this belief.

Since 1969, the industry has spent about $1.4 billion exploring in the frontiers. Exploratory wells in the Mackenzie Delta/Beaufort Sea can cost up to $25 million each. Our success ratios in the frontier areas have been about one in 34.

If, in 1968, we had deliberately rigged the casino, would we be playing in it with $25 million chips with a one in 34 chance of success?

I suggest that any reasonable person would have to conclude that we would not.

Does this statement really explain what happened? Does it do much more than lessen what even oilmen describe as *the* credibility gap between how they operate and how their many critics say they operate in Canada?

The oil industry in Canada didn't suffer unduly from a looming oil shortage. Much of the money from increased prices for gas and oil went into government coffers – about 17% to Ottawa and 43% to the producing provinces – but the petroleum industry did get roughly 40% of the price hikes. According to figures prepared by the Public Petroleum Association of Canada, in the period 1972-1975, Imperial, Gulf, Shell, Texaco (Canada) Ltd., and British Petroleum had net revenue increases ranging from 20.5% to 166%.* Yet higher profits do not necessarily lead to greater efforts to produce oil or to look for frontier sources. For instance, Imperial Oil's

* Laxer, James, and Martin, Anne, eds., *The Big Tough Expensive Job*, Erin (Ont.), Press Porcépic, 1976.

exploration development went from a total of $113 million in 1972 to $320 million in 1976, but $187 million of that was Syncrude and other heavy-oil extraction work, not frontier exploration. And David Lewis, former national leader of the New Democratic Party and now a professor of political science at Carleton University, Ottawa, has stated that the industry's exploration expenditures in the same period declined from 23% of revenue to 12%† (So much for the "adequate incentives." It will be very interesting to hear the excuses offered by Pirouetting Pierre and his fellow dilettantes as they keep raising the price of oil and gas. Why don't they take a leaf out of U.S. President Carter's book and impose a severe, escalating tax on gas-guzzling automobiles, a tax that could reach, as in the American case, $2,500.00 annually on some models by 1985? Or offer a tax credit to homeowners and businessmen who improve the insulation quality of their premises or install other, specific energy-saving devices?)

David Lewis also has something disturbing to say about natural gas supplies and a Mackenzie Valley pipeline.

> When you get to gas, the National Energy Board said in its latest report that Canada had about 24 years of supply. Even if it is only 20 years, there is sufficient time in which to develop new plans. There is no need to rush into the Mackenzie Valley pipeline tomorrow. There is time to consider the serious problems. As far as gas is concerned, the government has years during which to further the research work necessary to study the ecological effects; during which to arrange the environment in which the native peoples in the north have to live; during which to settle aboriginal claims. There is certainly nothing to justify irreversible decisions and actions before the Berger Commission tables its report.
>
> There is something else that astounds me. There are three organizations in Canada seeking pipelines: Canadian Arctic Gas Pipeline Limited; Foothills Pipelines Limited, also known as the Maple Leaf Line; and more recently, Polar Gas Limited,*

† *Ibid.*

* A consortium made up of TransCanada PipeLines (also a member of the Arctic Gas group), Canadian Pacific Investments, Panarctic Oils, Tenneco Oils and Minerals, Texas Eastern Transmission Corporation, and Pacific Lighting Gas Development Company.

which proposes to bring gas down from the eastern Arctic islands. I don't pretend to be an expert in the field and I can't give any technical advice, but I have read a good deal about these outfits and my common sense tells me that if I were the Minister of Energy, I would take a very good look at the Polar Gas proposal.

Why am I saying that? First, because Polar Gas has already found the gas. In the Mackenzie Delta only about four trillion cubic feet have been found. In the islands on which Polar Gas has explored and is working, they estimate that they have already discovered 17 or 18 trillion cubic feet. That is four times as much in gas reserves in the eastern Arctic as has been found in the Mackenzie Delta.

Secondly, there are very few, almost no native communities in the eastern Arctic, compared to the western Arctic. The disruption of the life of the native people would, therefore, be almost nil.

Thirdly, and perhaps most importantly, the pipeline must be totally Canadian. A pipeline in the eastern Arctic would not be anywhere near Alaska so the Americans couldn't play with it. They couldn't bring the gas from Prudhoe Bay and the gas in the eastern Arctic down together because the line would come down the side of Hudson Bay. The line would be Canadian.†

But the chances are that Polar Gas will not get the official nod ahead of a Mackenzie Valley or Yukon Territory pipeline. Why? Primarily because that bonanza of Prudhoe Bay gas has to be moved south, and the Exxon Corporation and other American oil companies have a large financial stake in its transportation to, and sale in, U.S. markets. Add the fact that far and away the major share of Canadian natural-gas production is controlled by the oil industry – meaning Exxon, Royal Dutch/Shell, Gulf Oil, British Petroleum, Texaco, Mobil Oil, and Standard Oil of California – because natural gas has long been, and still is, a by-product of oil exploration. Only one Canadian-controlled company (PanCanadian Petroleum) is among the "top ten" producers of natural gas, and it's No. 10. A Canadian journalist describes the power play very neatly:

All the resources known to exist in the Arctic, argues [F. K.]

† *Ibid.*

North [an eminent Canadian geologist], including the dry natural gas, "have no value whatever to this generation of Canadians. Canadians, as consumers, will not need to exploit any remote source of any of these materials during this century, unless we go on selling our easily-accessible resources to the United States. It is the Americans who are interested in resources from the Arctic, not Canadians. . . . This is the crux of the whole matter." United States industry needs the gas now. . . .

The [natural gas] lobby is a smooth one. Everything it does is smooth, from the promotional movies designed to scare Ontario householders about the coming energy shortage to the early commitment of $20 million for social, environmental and wildlife studies conducted under the control of the Arctic Gas Studies consortium. Its liaison with the politicians and civil servants of the federal government is also smooth. Under the management of William Wilder, an influential financier, and Vernon Horte, former president of TransCanada Pipelines, the Gas Arctic Group presented in the months before the Energy Board hearings on the pipeline an image of purposeful and unflustered preparation.

One reason for the emphasis on studies is that the Gas Arctic planners are anxious to avoid the blunders of the American lobby pushing for the trans-Alaska oil line down an 800-mile pipe to the ice-free southern Alaska port of Valdez, where the north slope oil will be loaded on tankers for delivery to the U.S. Pacific coast. That pipeline was stalled for years by conservationist appeals to the U.S. courts; Gas Arctic people cite it, therefore, as an instance of poor corporate planning and insensitivity.*

Another reason is that the Canadian oil- and gas-producing industry, like its counterparts in the Middle East or in South America, is obliged to meet the needs of its American owners: *their* aim is to maximize imports to, and profits in, the United States. Most of the rich petroleum being mined from the Athabasca tar sands by GCOS is being piped to American petrochemical plants, as will be Syncrude production. With the influence the petroleum lobby exerts on Ottawa, why should things be any different in the case of the subterra-

* Sykes, Philip, *Sellout: The Giveaway of Canada's Energy Resources*, Edmonton, Hurtig Publishers, 1973.

nean riches at Prudhoe Bay and in the Delta? And that lobby has been very, very successful in getting control of Canadian gas supplies. Whereas our oil is exported to the U.S. on the basis of 30-day licences, natural gas - now a very scarce commodity in the U.S. - is being piped across the border on the basis of long-range contracts that extend into the 1990s!

* * *

Bruce Willson, born and brought up in Canada, has seen the Canadian oil-and-gas industry from the inside as president of four U.S.-owned companies: Northwest Utilities Limited of Edmonton; Canadian Western Natural Gas Company Limited; Canadian Bechtel; and Union Gas. In 1974, he quit the industry. His principal reason for doing so is most revealing - and also makes very chilly reading.

> I have always looked on energy supplies as not being that different from water, and the environment, and electric power, and sewage lines, and roads; they're what you build an economy *upon*. The country can't stand the luxury of certain companies ripping off the consumer at prices related to what Arabian countries can charge an energy-depleted Europe rather than to cost. There used to be toll roads on Yonge Street [in Toronto] and that may have been acceptable, as long as what they charged was related to maintenance and cost of capital. But not if the tolls brought in ten times the cost of the road every year.
>
> It was fairly clear . . . that the government was being sold a bill of goods by the oil companies. By 1980, despite sizable imports of one trillion cubic feet annually from Canada, American supply was going to be only 60 per cent of estimated market requirements. Yet, U.S.-owned companies in Canada were asking to export more and more of our proven reserves and justifying this by speaking of vast potential reserves. Rather than the 53 trillion cubic feet of proven reserves, they were encouraging the government to think in terms of potential Canadian reserves of 725 trillion cubic feet. It was obvious they wanted the gas moved south as quickly as possible, to meet their rapidly growing deficiency.
>
> And very little thought was given to the cost of finding those 'potential' reserves and bringing them to market. The situation wasn't too difficult to understand. Between 1956 and 1970, proven reserves went up by about 7.5 per cent a year. Production

went up at a rate of 16.3 per cent. That's a collision course you don't need a doctorate in math to understand.*

* Laxer, James, and Martin, Anne, eds., *The Big Tough Expensive Job*, Erin (Ontario), Press Porcépic, 1976.

Chapter 6

People and Pipelines

Canadians and Americans grow up with the idea that Europeans discovered North America. They didn't. They merely mapped it. The point is well made in Charles M. Boland's book *They All Discovered America* (Doubleday, 1961). After discussing the first immigrants, who were followed by such assorted visitors as Phoenicians, Romans, a Buddhist priest from China, Irish monks, Vikings, and a Welsh prince and his followers, the author devotes the last chapter to Christopher Columbus. And it is thanks to Columbus that the first arrivals continue to be misnamed. Convinced that he had actually found the fabled East Indies, rich in spices and jewels, he called the brown-skinned, black-haired peoples he met "Indians."

The discoverers of North America were nomads who trailed herds of such well-fleshed, heavy-coated animals as caribou, musk ox, bison (buffalo), and woolly mammoth. Thousands of years ago, some of these nomads followed their sources of food and warmth out of easternmost Asia into what is now Alaska. At this time in the earth's history, much water was locked into the gigantic ice sheets – some of them a mile or more thick – that covered parts of the northern hemisphere. Ocean levels dropped considerably, and it would have been easy for animals and human beings to wander across the 60-mile sea bed of the modern Bering Strait, the short, shallow water gap where Asia and North America almost meet. Century after century, groups of hunting families drifted into North America, utterly unaware that they were leaving one continent for another. River by river, valley by valley – anywhere there was no impassable barrier of ice – they probed southward into the vast, continental interior. They gradually spread all over this New World, even as far south as Tierra Del Fuego at the tip of South America.

In time, the gigantic, North American ice sheet melted away, leaving in its sodden, stony wake millions of square miles of extra living space. Most of this had been so thoroughly ice-scoured that it is still little more than a rock-and-lichen wilderness, the practically treeless tundra of northeastern Canada and the naked, wind-whipped islands of the Canadian Arctic. But in the long valley of the Mackenzie River, where the land was less severely bulldozed by the ice, much of its surface is covered by bush and forest. Various groups of people filtered into this region.* The "People of the Small Knife," so-called by archaeologists from their preference for making small cutting tools out of flint or other hard stone, fished and hunted in the valley of the Mackenzie "proper." In the mixture of parklands and muskeg around Great Slave Lake lived the "Long Spear People," who shadowed a species of bison that had wandered north from the Great Plains. When not pursuing Wood buffalo, the Long Spears corralled some of the millions of caribou that migrated to and from the Arctic barrens each year. A second group of buffalo hunters, the "People of the Arrowhead," moved in around Lake Athabasca. And then there arrived in North America a group of comparative latecomers labelled by anthropologists the "Denbigh People." Named after an Alaskan cape where evidence of their bone and ivory harpoons has been found but better known as Eskimo, these particular people roamed Canada's Arctic coasts and islands. Two other similar cultures, the "Dorset People" and the "Thule People," have been traced in the north. The latter are thought to be the direct ancestors of the modern Eskimo or, as they call themselves, the Inuit.

One common denominator that links all these different

* One of the easiest "doorways" to the interior of North America is a little west of the Mackenzie Delta. Here, the 1,000-feet-high McDougall Pass is the lowest crossing place in the great barrier formed by the Mackenzie and Rocky Mountains. Thus, once in Alaska, the newcomers could ascend the Yukon River and one of its tributaries (the Porcupine), cross over via this pass, and enter the delta of the Mackenzie. The Mackenzie Valley, of course, stretches almost all the way to the edge of the Great Plains.

groups of human beings is the term they use to describe themselves. The names now used to identify Athapaskan-speaking Indians – Chipewyan, Yellowknife, Dogrib, Hare, Slave, Loucheux, Nahanni – are all names given by others, local or foreign. As Georges Erasmus, president of the Indian Brotherhood of the Northwest Territories, has explained, "All these names were imposed on us. We have always known who we were, particularly the old people. We have always called ourselves 'Dene' [pronounced Dennay]. Simply translated, we defined ourselves as 'people,' as different from the animals. With the coming of the Europeans, we developed the term 'Dene' to mean not only ourselves as a people separate from the animals, but ourselves as separate from the Europeans."* Similarly, the Eskimo call themselves Inuit, "The People." (Inuk is the singular term.) In their account of the Creation, they were the only people to spring directly from the earth, all others being the descendants of an Inuk woman and a dog. The word "Eskimo" originated with Indian tribes that sometimes came into contact with these Arctic nomads and means "eater of raw flesh." The Inuit, being a polite society, tolerate the misnomer.

Dene and Inuit became survival experts in an environment where it is impossible to grow crops or raise cattle and where the climate is harsh, often hostile. Without the benefits of agriculture and animal husbandry, metal and machines, they developed surerlative skills in hunting, trapping, and fishing. They fashioned everyday tools from stone, bone, antler, ivory, wood, and shell. They invented the birchbark canoe and the skin-covered kayak, the webbed snowshoe and the dog-drawn sled. Home was a hide lean-to, a domed snowhouse, or the insulation of a furred suit. In a landscape dominated by the sombre trees of the great northern forest or the dwarf bushes and tiny mosses of the tundra, they found ways to beautify their lives. Faces and bodies were embellished with intricate tattoo work. Red ochre and black graphite enhanced many a shirt or tent. Soapstone was worked into human and animal images. A favorite

* Canada. Department of Indian and Northern Affairs, "Transcript of the Mackenzie Valley Pipeline Inquiry," Ottawa, 1976.

horn spoon or a treasured bone knife was personalized by carving, dyeing, or bleaching. Even a birchbark pot could be improved by the judicious application of teeth marks.

Particularly among the Dene, survival was also a matter of intense group loyalty and a profound suspicion, if not hostility towards, anyone not a member of the group. Since a hunter often had to range several hundred square miles of land, water, or ice for a steady supply of food, the arrival of a stranger raised the spectre of starvation. Sickness or prolonged periods of bad weather were often blamed on the practice of witchcraft by someone outside the group. Wife-stealing was a common cause of inter-group quarrelling and fighting. It was often the reason that explained the sudden appearance of a stranger because, among the Dene, the strongest and best hunters kept several wives to process skins and backpack loads of meat from camping place to camping place. Once someone was killed in any such encounter, it set off a chain of fights and feuds. Sometimes these developed into small-scale wars, in which men wore armor of toughened hide or wood and fought with special arrows, spears, and clubs. As a general rule, however, all these northerners encountered few strangers in their lifetime. Most of a man's or a woman's contacts were the parent-grandparent family unit and an only slightly larger group of other relatives. In a world where sickness, accident, adverse weather, and cyclic declines in animal populations were the norm, you quickly learned the value of human interdependence and found ways to resolve differences of opinion. Sharing was all-important. The prime purpose of life was to help keep the group alive and well. When trouble arose, it had to be settled within the group, because, next to death, the worst punishment that could happen to anyone was to be banished from the group.

In the course of thousands of years, the Dene occupied today's Yukon Territory, Northwest Territories, northern Manitoba, Saskatchewan, and Alberta, and parts of northern and central British Columbia. Like other fellow immigrants to North America, their hunting-fishing economy left the land – and, for that matter, each other – unchanged. It is only within the last three hundred years that their lifestyles have been disrupted by successive invasions of outsiders. And the only appropriate term for these intruders is "foreigners."

This foreign influence was first felt in the watershed of the Mackenzie long before a Pedlar from Montreal or an HBC servant ever visited the region. For the better part of a century, those mercenary middlemen of the HBC fur trade, the Cree, ranged northern Canada, masters of the highly profitable trade in furs between the Dene and the far-distant, salt-water posts of the Company. If any Dene dared to challenge the Cree monopoly of trade with the Bay, he found himself looking down the barrel of a flintlock musket, a trade item that really revolutionized northern life. Expanding their lucrative commercial empire, the Cree harassed the Chipewyan and drove them deeper and deeper into the northern forest and farther and farther away from the Bay. Wave after wave of Cree intruders forced the Slave far north down the Mackenzie Valley and bundled the Sekani and Beaver so far west up the broad valley of the Peace that some of these tribes sought refuge in the Rocky Mountains. However, the Cree were by no means the only northerners to become actively aggressive as a result of European influences. The Chipewyan soon adopted the role of middleman and for many years dominated the Dene trade of the Mackenzie Valley. Around 1800, some Yellowknives in the Great Slave Lake area used guns to oppress the Slave, Hare, and Dogrib for the better part of the forty years. But, with the merger of the Hudson's Bay and North West companies in 1821, 150 years of increasing unrest and conflict gradually began to end. Unfortunately, the merger was also the beginnings of the end of Dene (and Inuit) independence.

Several accusations can be levelled at the fur trader. In the days of the Pedlars from Montreal, he was quite often a liar and a cheat. In the era of intense Nor'Wester-HBC rivalry, liquor flowed like water, and drunkenness and debauchery characterized many a trading session. But even a fair-minded Pedlar, Nor'Wester, or HBC employee could not anticipate the basic harm to which he was contributing in the course of his daily work: the steady destruction of an apparently endless supply of wildlife. String fish-nets, muzzle loaders and breech loaders, steel traps, and, eventually, the high-powered repeating rifle did the damage. The history of the fur trade is the story of a steady movement from one trapped-out region to the next untrapped one westward and northward. It is also the now-familiar Canadian story of the wanton exploitation of a resource until there's little or nothing left of it. The obvious example is the beaver. Around 1800, the North West Company alone was shipping about 100,000 beaver pelts out of Montreal. Sixty years later, the Hudson's Bay Company's annual auction sale in London could offer only 20,000. A less well-known example is the caribou. For centuries, this cousin of the reindeer was to the residents of taiga and tundra what the buffalo was to the Plains Indian: a life-sustaining source of food and clothing. And like the buffalo, a headcount of any one herd would have run into the millions. But the fur trade decimated the caribou. Once hunted simply to provide internal and external warmth, it ultimately became an easy source of bait and dog food. And, of course, traders needed huge annual supplies of caribou in the form of pemmican. As late as 1893, when J. B. Tyrrell of the Geological Survey of Canada explored and mapped the barrens west of Hudson Bay, he met several hundred Inuit who were truly "people of the deer." They lived exclusively on caribou. Everyone knows what happened to the buffalo, but how many people know that, by 1960, west of Hudson Bay, there were perhaps 200,000 caribou still alive in the whole country? How many know that, by 1960, no Indian or Inuit group was able to live off the barren-lands caribou on a year-round basis?

The next alien force was the Church. Several organizations have carried the word of God to Canada's northern peoples. But two have been particularly active in the valley of the Mackenzie: the Roman Catholic Oblate Order of Mary Immaculate (OMI), and the Church Missionary Society (CMS), Church of England in origin and now an arm of the Anglican Church of Canada. As early as 1862 there was a Roman Catholic diocese of Athabasca-Mackenzie centred on Fort Chipewyan and an Anglican church far downstream at Fort Simpson. And, just as the Mackenzie was once an arena of fur-trade competition, so it became the scene of another competition with, this time, Indian and Inuit families as prizes. Leapfrogging each other, Catholic and Protestant missionaries moved steadily downriver in pursuit of converts. To these, they offered both the consolation of religion and facilities that were not available from the trader or a disinterested government in Ottawa: education, and medical and nursing aid. Unfortunately, in retrospect, the story of the Church in the north is, like that of the fur trade, a mixture of blessing and bane.

Commerce and religion seem to be rival occupations. It is to a missionary's advantage to have a settled congregation in a village or town, whereas a trader wants a well-scattered population of trapping families. Trading is hindered, not helped, by a Church ban on Sunday work and travel, and the interruption of trapping and hunting routines by such special celebrations as Christmas and Easter. The employment of Dene and Inuit at Church establishments as hunters, laborers, carpenters and, under the Anglican system, as clergy, catechists, and lay readers, decreased the number of trappers and, therefore, reduced annual harvests of furs. Basically, the trader-trapper relationship was an impersonal give-and-take, whereas priest, nun, and pastor devoted themselves to the physical and spiritual well-being of their converts. However, despite these considerable differences, each group was remarkably similar – and foreign – in its thinking. Each sought to promote the values of "work" and "property" among peoples who appeared, to European eyes, lazy, obtuse, and culturally weak. Of the two, the Church has been the worst offender. In its zeal to convert, the Church came close to wiping out all vestiges of native cultural, religious, and economic values. Take, for instance, the matter of education.

Teaching the three R's to the children of HBC personnel, European or Métis, was always a problem in remote northern regions. The Hudson's Bay Company often engaged school teachers for its depots and encouraged missionaries to found schools at the larger posts. This was carefully considered policy, not paternalism. If some kind of education, academic or religious, was made available to children, then their parents were more likely to stay put. Even Governor

George Simpson grudgingly admitted of missionaries, Catholic or Protestant, that their presence was "a necessary evil for the benefit of the Indian population." In the long run, the natives had to be persuaded to give up their indigenous ways and become thoroughly productive in terms of selling furs and buying goods. However, in the case of the Church, what took place in the long run was a form of cultural genocide.

> . . . one of the most controversial areas of church involvement with the Native way-of-life is the educational system which was operated through residential schools and hostels until only a few years ago. Over the years, many thousands of Natives – Indians, Inuit, and Métis – have gone through the church-operated residential schools across Canada, and while they were once the pride of the churches who used them to recruit and finance large missionary outlays, they are now often considered a blot on the church's history in the North.
>
> Education has been a problem since the white man first arrived, and because it was always white-dominated it had – and still has – little relevance to most Natives. Statistics in 1974 reveal that across Canada, 94 percent of status Indian children* dropped out before finishing high school, as opposed to 12 percent for white young people. Less than 40 percent of Natives finish grade eight. While these abysmal statistics are not solely to be blamed on church residential schools, they illustrate the contempt that most white educators have had since the beginning of the assimilation process for anything Native. Curricula were designed by whites and taught by whites in an environment whose primary purpose was to make "apples" of all Natives – red on the outside, white on the inside. . . .
>
> The aim of the schools was assimilation. The children usually were forced to spend at least 10 months-a-year without seeing their parents, due to the high costs of transporting them from the centrally located school to the villages and settlements they called home. They had to learn to speak English and to regard the life their parents and relatives lived as dirty, pagan, and uncivilized. Even when residental school staff came in contact with those who could manage to visit parents, we were

* Under the Indian Act of 1876, revised in 1951, most Indians in Canada belong to registered bands and have a band number. In official terminology, they are Registered or Status Indians. All members of Indian bands that signed treaties are termed "Treaty Indians."

told there was an attitude of coldness that made them seem like intruders. . . .

The staff members of the schools varied. Many were dedicated if poorly trained and had, according to their own lights and the strict rules laid down for them by the churches, the best interests of their charges at heart. But the sisters, lay brothers, and priests of the Roman Catholic church and, to an even greater degree (because they were often more inflexible and pietistic), the lay men and women and priests of the Anglican Church, dealt with their students as "charges" or wards in a most autocratic and paternalistic fashion.

The residential schools were run efficiently and with great frugality, since until late in their existence they were supported almost entirely from church funds. Only in 1958 was considerable government money put into their operation, and by 1969 church contacts with the government had been terminated except in the Northwest Territories. . . .

Former employees of the church-school system, regardless of denominational affiliation, tend to attribute complaints by their former pupils to a lack of gratitude for what was done. "Sure we made mistakes and there were failures. Perhaps we were too strict. But if the churches hadn't gone in and brought those kids out to school, the very leaders of the Native organizations today wouldn't be able to read or write. They never acknowledge that," a former administrator pointed out to us.

But most of the "graduates" we talked with said the standard of instruction was in many cases so low that few people ever reached an academic level that enabled them to enter high school. It is noteworthy that most of the few university graduates and professional people have come from the post-war years when students started to go to schools in the communities and more-and-more of the church schools became only residences.*

One graduate had something very personal to say of this school system in August, 1976, when testifying at one of the last community hearings held by the Mackenzie Valley Pipeline Inquiry. Born at Fort Franklin on Great Bear Lake, Gina Blondin was sent at the age of six to the village of Camsell on the Mackenzie River to be treated for tuber-

---

* McCullum, Hugh and Karmel, *This Land Is Not For Sale*, Toronto, Anglican Book Centre, 1975, pp. 175-177.

culosis. When released from hospital, she was lodged in a hostel because both her parents were sick with tuberculosis. "When I left Fort Franklin, I did not know a word of English. After five years of hostel life, I did not know a word of Slavey. I will always find the loss of my language one of the most tragic and destructive things that education has done for me."

A third, alien force to disrupt the world of the Mackenzie Dene and Inuit is the colonial-type government imposed on them by Ottawa.

As recently as 1954, the Yukon Territory and the Northwest Territories – that is, the third of Canada that sprawls north of the sixtieth line of latitude and the Province of Quebec – were administered from Ottawa in "an almost continuous state of absence of mind." This was the accurate description used by Prime Minister Louis St. Laurent when, in 1953, he placed before the House of Commons a bill proposing the establishment of a Department to administer and develop the Canadian North. North of sixty and of Quebec's arctic shoreline, the only effective Canadian authority was an unlikely trinity composed of the Anglican and Roman Catholic churches, the Hudson's Bay Company, and the Royal Canadian Mounted Police.* But, by 1953 it was high time to pay serious attention to the nation's immense northland. For one thing, the United States wanted to construct the DEW Line, a series of distant-early-warning radar outposts in the Arctic that would detect any Soviet air attack across the North Pole. This immediately roused fears that Americans would assert ownership in the Canadian northland as they had tried to do at the time of the Klondike gold rush in 1898. There were, too, disturbing stories circulating within and without Canada about the misfortunes of its northern peoples. It was bad enough that wildlife had been dangerously thinned out in many places "north of 60." Worse still, in the early 1950s, many Dene and Inuit were suffering

* Between 1913 and 1954, the three Districts of Mackenzie, Keewatin, and Franklin were officially governed by the Deputy Minister of the Department of the Interior (later the Department of Mines and Resources) and a council consisting of representatives of the Anglican and Roman Catholic missions, the Fur Trade Commissioner of the HBC, and the Commissioner of the RCMP.

from illnesses never known before the arrival of foreigners: smallpox, measles, influenza, diphtheria, typhoid fever, venereal disease. In particular, tuberculosis was showing a definite upswing. Thus, in 1954, Ottawa hastily created the Department of Northern Affairs and National Resources.

Today, the Department's many critics say that the end product is an exercise in true, no-expense-spared bureaucracy. In 1954-1955, various federal agencies, which employed a total of 3,000 people, serviced the Yukon and Northwest Territories at a cost of some $20 million. Barely twenty-five years later, 15,000 federal civil servants of various kinds were administering the lives of some 65,000* northerners at a cost of close to $1 billion. Home base for the "Feds," as they are referred to in the Yukon and in the Northwest Territories, is Centennial Towers in downtown Ottawa. From here, the Minister of the Department of Indian Affairs and Northern Development (DIAND) rules 30% of Canada in a fashion similar to that in which the Family Compact ran Upper Canada (Ontario) and the Chateau Clique ran Lower Canada (Quebec) about a hundred and fifty years ago. The Minister *appoints* as his personal representative a civil servant, resident in Whitehorse, as Commissioner of the Yukon Territory, and another, resident in Yellowknife, as Commissioner of the Northwest Territories. Via the authority of the Minister, these heads of government have the power to veto the actions and recommendations of a legislative council elected by Territorial residents. In other words, northerners have only a part say in the running of their own affairs. However, when it comes to money matters, the federal government's master-servant relationship with the territorial governments (whose civil servants wryly describe themselves as the "Terrifieds") is complete. Unlike a province, which has control of its own natural resources, surface and sub-surface land rights in the Yukon Territory

* At the last census count, the population of the Northwest Territories was approximately 42,000. Of this total, about 13,000 are Inuit, perhaps 7,500 are Dene (exact figures don't seem to be available), and possibly 8,000 Métis. To complicate matters, only a fraction of the remaining 13,000 or 14,000 of the Territorial population that are non-native are permanent residents.

and in the Northwest Territories are owned by Ottawa and are controlled by DIAND. Unlike a province, which has the constitutional right to raise and spend its own monies, territorial governments can borrow money only with Ottawa's permission. In any case, these governments are overwhelmingly financed in their operations by Ottawa with funds extracted from the pockets of Canadian taxpayers.

Government – any government in any place at any time – tends to be inefficient and wasteful. DIAND, of course, is no exception. It has a well-established record as the nursery of good ideas and the graveyard of bad ones. It has attempted and failed to promote just about everything from folk festivals to yak breeding. Since 1954, contrary statements and confused reports have highlighted the Department's activities. The most baffling reports of all are probably those emphasizing that the more that government money is spent in the Canadian North, the less its native peoples benefit. Despite re-location in permanent settlements, mostly in subsidized-rent housing built with DIAND funds, despite the provision of schooling, most Indians and Inuit have low incomes. In 1968, for instance, they were earning, on average, about $500 a year while other northerners were averaging close to $3,000 (at that time, 70% above the national average). How come?

The factors are varied and complex, but the answer always seems to boil down to the same thing. Ottawa has long been much less interested in "Indian Affairs" than it has been in "Northern Development." In fact, from the very beginnings of settlement in western Canada and mineral exploration in northern Canada, Indian and Inuit have been an embarrassment to their official guardian, the federal government. In 1870, writing to the Lieutenant-Governor of the newly created Province of Manitoba – who had also been made Lieutenant-Governor of the "North-Western Territory,"* until that year the exclusive domain of the Hudson's Bay Company – the official line was laid down in blunt terms

* This was the terminology first used by the government of Canada. A few years later, the name was changed to "North-West Territories". Later still, it became "Northwest Territories."

by Edmund A. Meredith, Under Secretary of State for the Provinces:

> You will also turn your attention promptly to the condition of the country outside of the Province of Manitoba, on the North and West; and while assuring the Indians of your desire to establish friendly relations with them, you will ascertain and report to His Excellency [the Governor General of Canada] the course you may think the most advisable to pursue, whether by Treaty or otherwise, for the removal of any obstructions that might be presented to the flow of population into the fertile lands that lie between Manitoba and the Rocky Mountains.*

By means of treaties, the "obstructions" were removed within the space of seven years to make way for the settlement of the Canadian prairies between Winnipeg and the Rocky Mountains and between the 49th parallel and the great forest that edges the prairies to the north.

Canadians have often deplored the American treatment of native peoples while praising their own, yet the difference is one of method, not of intent. In 1871, Washington ceased making treaties with Indians and used the U.S. Army to take over new territory and subjugate its native residents. In Canada, between 1871 and 1877 for example, seven agreements made with prairie Indians secured their hunting grounds, rivers, and lakes in return for inexpensive presents of various kinds and quite nominal annuities of money. In the name of the monarch of Canada, Ottawa employed treaties in order to, as the law puts it, "extinguish" all rights to land ownership. Small groups of officials called "commissioners" travelled a treaty region by horse or canoe to meet Indian bands. With the aid of interpreters, usually Métis, missionaries, or traders, the commissioners explained Queen Victoria's need of land for farming, lumbering, mining, etc. The general Indian impression of the whole thing was that "the Great White Mother" was asking for two things: their loyalty, and the *use* of large tracts of their land. Puzzled though they were by the promise that they would be paid for both, Indians who discussed the treaties with the commissioners were, above all else, concerned about their rights

* Public Archives of Canada, *Sessional Papers*, 1871, No. 20.

to hunt, trap, and fish. As a result, these rights were guaranteed (subject to such later changes as Ottawa might see fit to make). Payment, in money and land, sealed the transaction – at least it did in the minds of those Europeans who signed the treaties. Basically, the deal included the payment of $5.00 a year to every Indian man, woman, and child, the grant of 1 square mile per family of 5, and the supply of twine for nets or ammunition for hunting. (The annual payment is still $5.00 per person, per year, although headmen and chiefs receive slightly more.)

So the "obstructions" to settlement on the prairies were removed and these lands exploited peaceably – and not a little profitably. As for territory lying north of the prairies, land unsuited to farming or ranching, treaties with its Indian owners could wait. Despite appeals from HBC officers and missionaries in times of want and famine, Ottawa ignored its many other Indian wards. In the years 1870-1889, the government's attitude was "no treaty, no help." The prime minister himself agreed with this policy. Sir John A. Macdonald noted in the margin of a letter sent in 1884 by the Lieutenant-Governor of the Territories to the Superintendent General of Indian Affairs, ". . . the making of a treaty may be postponed for some years, or until there is a likelihood of the country being requested for settlement purposes."* Then, between 1899 and 1921, two treaties were signed with northern Indians. In each case, there was a strong economic motivation for Ottawa to do something about establishing its ownership of land in the drainage basin of the Mackenzie River.

In the late 1880s, Ottawa was aroused by repeated reports that the entire Mackenzie Valley appeared to be floating on oil. Members of the Geological Survey of Canada had twice detected the presence of potentially enormous deposits of oil and natural gas in the "Athabasca country," that historic Eldorado of the fur-trade surrounding Lake Athabasca. Then, exploring steadily downstream, veteran surveyor R. G. McConnell reported that,

The Devonian rocks throughout the Mackenzie valley are

* Public Archives of Canada, RG 10, BS, file 241, 209-1.

> nearly everywhere more or less petroliferous and over large areas afford promising indications of the presence of oil in workable quantities. . . . [The valley's] remoteness from the present centres of population, and its situation north of the still unworked Athabasca and Peace River oil field will probably delay its development for some years to come, but this is only a question of time. The oil fields of Pennsylvania and at Baku [in Persia] already show signs of exhaustion, and as they decline the oil field of northern Canada will have a corresponding rise in value.*

McConnell's description of an "oil field of northern Canada" triggered a Senatorial committee of enquiry that mentioned, among other things, the presence in the Mackenzie basin of "auriferous deposits of silver, copper, iron, asphaltum, and other minerals." Almost overnight, the North-West Territories became, in Ottawa's thinking, the "Petroliferous and Auriferous Territories," and plans were soon made to initiate treaty negotiations with the approximately 16,000 native inhabitants of what were then called the Peace River, Athabaska, and Mackenzie River districts.

For various reasons, no treaty was concluded until 1899. By this time, the Klondike Gold rush was in full swing, and a spin-off activity was the invasion of the Great Slave Lake region by prospectors seeking gold and any other precious metal that could be found there. All this activity produced frightening change. The first Royal North West Mounted Police patrols into the North-West Territories occurred in 1897, and successive Police patrols reported a growing traffic of illegal liquor, the widespread use of poisoned bait by European trappers and hunters, and the obvious disregard of trappers and miners alike for Indian territorial and property rights. In 1898, Ottawa and Winnipeg newspapers reported that Indians encamped at Fort St. John in the upper valley of the Peace River refused to let miners or the Police go farther north until a treaty had been signed. Their horses were being stolen, and they also feared that fur-bearing animals would be driven out of the surrounding countryside.

* Geological and Natural History Survey of Canada, *Report on an Exploration in the Yukon and the Mackenzie Basins, N.W.T.*, Annual Report, 1888-1889.

In the Athabasca region, not only were horses being stolen but sled dogs shot and bear traps smashed. Thus, there was reason enough for government to step in, acquire title to land, and enforce law and order. But another, underlying consideration influenced the treaty-making, one that made the welfare of Indian and Métis a secondary consideration.

Ottawa's reasoning was simple. The direct route to the Klondike overland through the Peace River country, or the much more indirect one via the Athabasca, Slave, Mackenzie, and Liard rivers, had to be kept trouble-free. The Territory of the Yukon had just been created (1898), the Mounted Police had firm charge of that region, and the gold was flowing south in comforting quantities. In addition, as Amédée Emmanuel Forget, the Indian Commissioner of the Northwest Territories, noted in 1898, there was "considerable activity in mining matters in the valleys of the Peace and Athabasca rivers north of the Treaty 6 area and also in the upper valleys of the Peace and Liard rivers." Thus, beyond these points, Forget argued, "I do not consider that the Government would be justified in undertaking the negotiation of treaties, which would involve very heavy outlay for comparatively inadequate returns insofar as the value of the territory to be ceded, or the rights of the Indian owners, are concerned. . . ." A few months later, again writing to the Indian Affair's Secretary, Forget emphasized that the treaty should not include the districts inhabited by "the Indians about Great Slave and Great Bear Lakes and along the Mackenzie River . . . their territory so far as it is at present known is of no particular value and they very rarely come into contact with Whites."* Years later, confirmation came from David Laird, an Indian Commissioner working out of Winnipeg at the time of the treaty. In a letter to the Indian Affairs Secretary, Laird explained the boundary decisions as a means "to protect and control whites who were going into the country as traders, travellers to the Klondike, explorers and miners."†

In the summers of 1899 and 1900, at various locations, the

---

* Public Archives of Canada, RG 10, BS, file 75, 236-1.

† *Ibid.*

Dominion of Canada obtained yet another excellent land bargain. It was made between Her Most Gracious Majesty the Queen and the Cree, Beaver, Chipewyan, "and other Indians, inhabitants of the territory within the limits hereinafter defined and described." In return for the ownership of their lands, they were guaranteed (with some qualification) hunting, trapping, and fishing rights and annual payments of $25 to each chief, $15 to a headman (not to exceed four such to a large band), and $5 to every other Indian.* Once again, assurances by missionaries and Mounted Police encouraged reluctant Indians to trust the motives and promises of Her Majesty's representatives.

Much the same story explains Treaty 11, made in 1921. A year earlier, on Indian territory not surrendered by treaty, the first great Canadian oil find of the twentieth century came gushing out of the ground at Norman Wells, far down the Mackenzie's mainstream. Imperial Oil had found its first northern bonanza. Every time the well was uncapped, a 70-foot column of "black gold" shot up into the air, saturating the surrounding bush and muskeg with excellent quality crude oil. It was, Canadian newspapers claimed, the "Biggest Oilfield in the World" and stretched at least some 400 miles downstream to the Arctic Ocean. According to the *Sydney Post*, the first well's production was spouting one thousand gallons an hour, and Imperial Oil was preparing to punch a district "200 miles in width and nobody knows how long full of holes for wells." (Nobody paid any attention to the knowledgeable opinion of Imperial Oil's Calgary manager. M. B. Green was quoted in the *Calgary Herald* as saying ". . . the oil there is not and will not be worth one red cent, for it is doubtful if a railroad, the only means by which the oil could be transported upon a commercial basis, would even be built to that far northern outpost; certainly not unless oil should be found there in overwhelming quantities. . . . Whether it is there in sufficient quantity to warrant the spending of millions of dollars to attempt to get it out is not known yet by any means.")

* There were also various handouts in the form of silver medals, flags, a suit of clothing for chiefs and headmen, salaries for such schoolteachers as might be appointed, axes, handsaws, and various agricultural tools, animals, and seed supplies.

The implications of the oil find galvanized officialdom into sudden activity. A North-West Territories' government had been legislated by Ottawa in 1875, but it was the news of Norman Wells that caused brand-new councillors to be appointed (not surprisingly, all of them federal civil servants, and all of them resident in Ottawa) and an administration organized at Fort Smith on the Slave River. The main items on that government's first agenda were oil and gas, and within a month a Mining Recorder's Office had been set up and staffed at Fort Smith. Ottawa, of course, was ecstatic. Arthur Meighen of the Conservative Party, Leader of the Opposition in the Commons, told the House that "The eyes of the mineral operators are upon us today. Oil is King." Another MP demanded federal ownership of this "vast domain of country, rich in natural resources and favorable for development." And, as it happened, the Department of Indian Affairs established federal ownership suspiciously quickly. For years, it had refused the pleas of one of its senior, on-the-spot officials and those of several church missionaries to make treaty with Mackenzie River Indians in order to protect them from periodic sickness, starvation, and, in some cases, destitution. However, within six months of the oil find at Norman Wells, the Department was advising the government that "immediate action should be taken to provide for the entrance of the Mackenzie River Indians into Treaty," and treaty costs had been included in the Indian Affairs budgetary estimates for 1921. The reasons for this, according to the Department, were "The early development of this territory is anticipated and it is advisable to follow the usual policy and obtain from the Indians cession of their aboriginal title."*

Inspector Henry A. Conroy, the official who had long argued that the natives needed Ottawa's money, medical aid, and legal protection, was named Commissioner "to negotiate a treaty with the Indians occupying the territory north of the sixtieth parallel and along the Mackenzie River to the Arctic Ocean." Conroy was handed a copy of the treaty

* Public Archives of Canada, RG 10, BS, file 336877, Superintendent General of Indian Affairs to the Governor General in Council, 9 February, 1921.

and warned to make no promises of any kind extra to those embodied in the treaty. He was ordered to obtain the consent of the native peoples via their chiefs and headmen at nine locations on Great Slave Lake and on the Mackenzie River.

Two people of very different backgrounds have passed their own judgments on Treaty 11. One is Father René Fumoleau, of Yellowknife, N.W.T. Since 1952, he has served as a missionary of the Oblate Mission of Mary Immaculate (OMI) at various places on Mackenzie waters. In 1973, at the request of some Dene friends, he began preparing a few notes on the history of the 1899 and 1921 treaties. Disturbed by the strong economic impetus that seemed to underlie treaty making, Father Fumoleau's initial investigations turned into a two-year search in government, church, and HBC archives and in communities up and down the Mackenzie to pull together the story of Ottawa's dealings with the Dene and their memory of these dealings. The result was a 400-page book that maintains the treaties were unjust, were never explained, and are even more unjust today. After describing the background to Treaties 8 and 11 and examining the motives of those who drew up the treaties and those who signed them, Father Fumoleau is convinced that, as far as the Dene were concerned, no cession of their land was involved. As he says in his summary of the Treaty 11 negotiation,

> When the Indians agreed to sign Treaty 11, they were promising their friendship. For over 100 years the Northern Indians had been able to help newcomers to their country. They welcomed and assisted the white man and trained him to survive and adjust to northern conditions. By 1920 neither the Indians nor their assistance were needed. Rather, the Indians had to depend on the white man's treaty as his way of offering them his help and friendship.*

Mr. Justice William G. Morrow of the Supreme Court of the Northwest Territories has come to the same conclusion. In his Court, in March, 1973, sixteen Dene chiefs filed a

* Fumoleau, René, *As Long As This Land Shall Last: A History of Treaty 8 and Treaty 11, 1870-1939*, Toronto, McClelland and Stewart Limited, 1975.

caveat (legal warning of a claim of prior interest) to ownership of about 400,000 square miles of land in the western portion of the Territories. After listening to almost six months of evidence – some of which he investigated for himself in native communities along the Mackenzie River – Justice Morrow declared himself satisfied that these chiefs had a right to file claim to the lands in question and that the government of Canada had "a clear constitutional obligation to protect the legal rights of the indigenous people in the area." . . . It was his considered judgment that, despite the wording of Treaties 8 and 11, "there is sufficient doubt on the facts that aboriginal title [to the ownership of lands and waters] was extinguished." His opinion, however, didn't go unchallenged.

> In June, 1973, one of the few representatives of white authority generally respected by the peoples north of sixty was the object of an unprecedented attack by the Trudeau government. It filed with the Federal Court of Canada an application to remove the issue of the frozen land titles from Mr. Justice William Morrow of the Northwest Territories Supreme Court, a living symbol of Canadian justice in the North. Arguing that the question of land title in the Territories was out of his jurisdiction, the Trudeau government wanted the freeze lifted through the use of an Ottawa-based court of equal constitutional standing to Morrow's own. Mr. Justice Morrow reasserted his authority until such time as it was rejected by the Territorial Court of Appeal.
>
> In a twenty-nine-page judgment, Judge Morrow declared the federal move "the first time in the history of Canadian jurisprudence, the first time since Confederation, that one superior court has been placed under attack by another superior court judge of equal status." It was his duty, the northern judge said, "to let the people within my jurisdiction and their Parliament know what is happening, so they will have their judge's side of the story. . . . The federal government is undoubtedly anxious about these proceedings, having such a substantial property interest in the lands involved – but then, so do the Indians."*

* Sykes, Philip, *Sellout: The Giveaway of Canada's Energy Resources*, Edmonton, Hurtig Publishers, 1973.

Ottawa finally accepted the fact that Justice Morrow had placed a freeze, however temporary, on all land transactions within those 400,000 square miles of land. "The Feds" gave ground on the matter, albeit very grudgingly. In January, 1974, the Department of Indian Affairs and Northern Development announced the formation of a committee that would investigate the means by which a comprehensive claims settlement of lands in the Northwest Territories could be effected. But DIAND refused to arrange a land settlement *before* pipeline construction began. Early that same year, the Minister of the Department, Jean Chrétien, laid it on the line:

> This government, after weighing all the factors involved very carefully, has come to the conclusion that a gas pipeline down the Mackenzie Valley is in the national interest. We will safeguard the rights of the people and we will protect the environment, but we will build the pipeline.*

The entire treaty-making process - which began with a Royal Proclamation in 1763 and covers a period of time extending to 1923 - is suspect. It is not treaty making in the normal sense of the term whereby two sovereign states negotiate and sign a series of terms of agreement acceptable to each. Indians were never considered independent and self-governing in the same way that nations regard each other as sovereign states. While the Indians' right to own land was respected, they were always regarded as being subjects of the Crown. Treaty terms were imposed, not negotiated. Had Indians refused the terms, their land would have been taken from them anyway. Yet it is extremely doubtful that, when they made their mark on a piece of paper, Indians understood they were quite literally handing over their ancestral lands to others. To an Indian or a Métis (and, for that matter, to the Inuit) the ownership of a dog, a sled, hunting equipment, or cooking utensils is one thing, but the earth belongs to no one individual or group. A family, band, or tribe may have long-established, clearly-recognized

* Dosman, Edgar J., *The National Interest: The Politics of Northern Development 1968-75*, Toronto, McClelland and Stewart Limited, 1975.

hunting and fishing rights in this, that or the next piece of territory, but the exclusive, *personal* possession of land is totally foreign to their thought and experience. Land is a perpetual inheritance. Each man and woman has the use of it, but has no right to destroy it in the course of their daily life and thus deny its benefits to successive generations. Hugh and Karmel McCullum, who have spent much time in the Canadian North as a project team working on behalf of the Anglican, United, and Roman Catholic churches, have talked with hundreds of native people in the course of thousands of miles of travel. The McCullums were told, time and time again, "without the land people have no soul, no life, no identity, no purpose."

> We are people of the land. We love the land but the land is no longer what it was.
>
> We see the land as having been taken into the white man's society, his economy. It is covered in asphalt, surveyed, scarred, tracked in the search for oil and natural gas and minerals that lie under it.
>
> Sadly we know that we cannot use it much in the old ways. The animals are hiding or dead, the fish are poisoned, the birds fly away. So we must seek a way to live in this new way, but we must not sell our land or allow it to be taken away. It is only for our use, even if the use is a new use. The land is for our children, not for sale. The land is still part of us and we are part of it.*

It is this radical difference in outlook and life style that is at the root of the conflict between the native peoples and officialdom over *any* pipeline.

* * *

The proposed Mackenzie Valley pipeline has been billed as the single largest job of construction even conceived by man. Whether that's true or not, its statistics are staggeringly impressive, even to Canadians and Americans accustomed to bigger-is-better facts and figures. Two business groups, Canadian Arctic Gas Pipeline Limited (CAGPL) and Foothills

* McCullum, Hugh and Karmel, *This Land Is Not For Sale*, Toronto, Anglican Book Centre, 1975, pp. 9-10.

Pipe Lines Limited, compete for permission to build a pipeline following the grant of a right-of-way by Ottawa. According to CAGPL, their costs will be in the $7-10 billion range. Running 2,400 miles south from Prudhoe Bay to link up with existing transmission lines in Alberta, this particular line will carry enough natural gas daily to heat 35,000 homes for a whole year. Its construction will require, among other things, at least a million tons of piping that can tolerate temperatures ranging from 50 below to 90 above, the technical know-how to cross over or under some 600 rivers and lakes, and time-tabling the energy and efforts of 6-7,000 workers.

CAGPL publicists describe their pipeline as "a thread running across a football field," which is about as accurate a statement as saying that someone's adolescent sister or daughter is "slightly pregnant." In point of fact, the land-use requirements of a Mackenzie Valley gas pipeline are enormous. In the Delta, gas-gathering systems will have to be laid down and gas refrigeration plants built. In many places, pipeline construction will involve a road bed of gravel several yards wide and several feet deep to support above-ground sections of pipe and its pylons. Compressor stations will have to be built to keep the chilled gas pressurized and moving in a controlled flow. A complex, computer-operated system will have to be installed to monitor the flow of gas. In order to provide electricity for the compressor stations, the computer system, and its supporting microwave communications network, power-generating plants will have to be installed on river systems.

Then there's the problem of moving labor and supplies from place to place within the Valley. The Mackenzie River is already a transportation corridor for a considerable amount of summer transport by water and year-round haulage by air. Every May-September, the Northern Transportation Company Limited (NTC), a federal-government agency, operates a fleet of tugs and barges out of Fort McMurray on the Athabasca River and the hub terminal of Hay River at the western end of Great Slave Lake. To communities along the Mackenzie's mainstream, on the north shore of Alaska, and on some Arctic islands, NTC carries just about everything from a wedding dress in a cardboard box to a 50-ton piece of machinery in a rough, wooden crate. And roaring out of Edmonton's International Airport every day of the week, Pacific Western Airlines jets service Fort Smith, Hay River, Yellowknife, Fort Simpson, Norman Wells, and Inuvik. Despite these facilities, the most important, inhibiting effect on natural-resource development in the N.W.T. is transportation, which basically provides domestic supplies to hamlets, villages, and towns whose economy is based on hunting, fishing, and government services. There is no railway beyond Great Slave Lake. There is no road, let alone a highway, in the 1,000-mile stretch of river between Fort Simpson and the Delta. Yet, as with the Alaska oil line, masses of steel piping will have to be trucked into the Valley in winter, the peak period of construction in northern latitudes. Pipeliners working out of trailer camps will be using hundreds of tractors, bulldozers, cranes, ditch-digging and line-welding machines that will need storage and servicing facilities, all of which postulates a highway and some sort of access-road network. New wharves will have to be built to accommodate a doubled or trebled barge traffic. Air strips and helipads will also have to be constructed.

But what happens if large oil strikes are also made in the Delta? So far, petroleum companies have found only natural gas in large quantities, but the lower Mackenzie region bears all the geological earmarks of a major petroleum basin. At any time, an oil strike as great as that made next door in northern Alaska could come gushing out of the ground – or spewing into the shallow waters of the Beaufort Sea. This would require the construction of an oil pipeline. In addition, the possibility of a Mackenzie Valley rail line has been investigated, and its supporters say that it's a perfectly feasible alternative to an oil pipeline. So where will all the demands for land and right-of-ways end?

The key to a northern pipeline is land, land on which live Inuit – with whom Ottawa has never made a treaty – Dene, and Métis. And they don't relish the prospect of being inundated, probably drowned, by a massive flood of foreigners. They will be the last to benefit from the exploitation of oil and gas. As a result of DIAND's amazingly generous royalty concessions to petroleum companies operating in the North, the government share of the potential gains is negligible. (When production starts, royalties are 5% for the first three years – barely a third of what is usually charged oil companies in the U.S. – and 10% thereafter.) In Alaska, a natives-claim settlement has set aside 40 million acres of land, and guaranteed $462 million in federal funds and $500 million in future oil royalties to 60,000 Indians, Inuit, and Aleuts. Those in the N.W.T. and the Yukon are due to receive nothing for their land and nothing for its subsurface riches. And the experience of the Alaskan oil pipeline suggests that native peoples will get only a tiny fraction of the construction jobs available. As in Alaska, the labor force will be experienced construction workers from the south. The peoples of Canada's "two colonies" think they'll be lucky to get the "pick and shovel work," which they say is all they got out of past developments. "Anyway," they ask, "what good are pipeline jobs? What will become of the land and the animals? Won't they be destroyed by the construction?"

There are no easy answers to these questions. But one man has made an exhaustive investigation of the problems and proposed certain solutions. He is the man that Dene, Métis, and Inuit refer to with respect and affection as "our Judge."

Thomas Rodney Berger is no stranger to the issue of native rights. The son of an RCMP officer, part of his boyhood in western Canada was spent on an Indian reserve. A few years after graduating from the University of British Columbia, he was legal counsel for Métis and Indian trappers in their dispute with BC Hydro over muskrat-hunting rights in the delta of the Athabasca River. In 1969, when the Nishga Indians of northwestern British Columbia sued the provincial government for non-recognition of their right to ownership of the Nass River valley, they chose Berger as their lawyer. Indeed, he has long been known as a champion of the underdog. An adversary of the federal government in court and a one-time Opposition–New Democratic Party–member of Parliament, thirty-eight-year-old Tom Berger was surprised when he was appointed to the British Columbia Supreme Court in 1971 by a Liberal minister of justice in Ottawa. But life is nothing if not full of Liberal surprises.

Three years later, Mr. Justice Thomas Berger was appointed by Jean Chrétien, Minister of Indian Affairs and Northern Development, to head a Commission of Inquiry into the social, environmental, and economic impact of a natural gas pipeline upon the Mackenzie Valley.

The judge's initial response to his task was characteristic. He took his wife and kids north that summer. By jet, bush plane, helicopter, and canoe, the Bergers visited communities up and down the Mackenzie River and also in the Yukon Territory. It was a 10,000-mile trip that took them to almost every settlement that was liable to be affected in some way or other by a pipeline. The judge wanted to get to know something about the people and chatted, ate, and sipped tea with some of those who, as it turned out, would appear before his Commission to testify. It was quite an experience, and he later told a reporter,

> I remember walking the streets of an Eskimo village early one morning in July when it was bright as day at two o'clock. Everyone was up and about. We met an Eskimo and he invited us in for tea and we must have stayed there drinking tea and talking about the north until five o'clock. He didn't know who I was. He just talked about his children, about hunting in the bush. He showed me around his house.
>
> The judge found the same openness in other places, though not always with tea and not always for hours of talk. "One thing about the native people is that a pause in a conversation doesn't disturb them," he said. "They expect gaps in conversation. If you can't think of something to say you don't say it. You often just sit there looking at the river."*

By the fall of 1974, there was still no word from Mr. Justice Berger as to when the Commission's formal hearings would begin. There had been preliminary hearings in the spring in Yellowknife, Inuvik, Whitehorse, and Ottawa, and again in Yellowknife in September. During these, CAGPL had made a strong, sustained pitch to get the pipeline contract and had won support from certain northern businessmen and

---

* O'Malley, Martin, ***The Past and Future Land: An account of the Berger Inquiry into the Mackenzie Valley pipeline***, Toronto, Peter Martin Associates Limited, 1976.

from the Northwest Territories Chamber of Commerce. But equally strong opposition had been voiced by what the media and others termed "parka politicians," the young, angry leaders of such native organizations as the Indian Brotherhood of the Northwest Territories, the Committee for Original Peoples Entitlement (COPE),* and the Métis Association of the Northwest Territories. They were practically penniless and needed money and time with which to research and present a counter proposal on behalf of their peoples. Late in 1974, these organizations and some environmental groups received federal funding to prepare their presentations to the Mackenzie Valley Pipeline Inquiry.

In the Territorial capital of Yellowknife on March 3, 1975, in the glare of the arc lights of TV camera crews working in the ballroom of the Explorer Hotel, Mr. Justice Berger tapped a gavel to open his investigation of a host of pipeline-related matters. Gazing through steel-rimmed spectacles at the various lawyers, petroleum executives, native-rights spokesmen, government officials, and media people assembled before him, his opening lines set the tone of his chairmanship for the next twenty-one months. "Ladies and gentlemen, we are embarked on a consideration of the future of a great river valley and its people. This Inquiry is a study whose magnitude is without precedent in this country. I have been guided by the conviction that this Inquiry must be fair and it must be complete. We have to do it right." The very next day, in the House of Commons in Ottawa, a comment by a Liberal cabinet minister underlined the difficulty of Berger's task, which was not to decide if a pipeline should or should not be built but *how* it would be built. Acting Prime Minister Mitchell Sharp, declining to promise that the Trudeau government would wait for the results of the Inquiry before giving a go-ahead on the pipeline, remarked "we fully expect that Mr. Justice Berger will be able to complete all his hearings in time for us to take action." The response from Yellowknife was quick and to the point. "My mandate is to conduct a fair and thorough inquiry. That

* COPE has its headquarters in Inuvik. It represents the Inuit of the Western Arctic and is a regional affiliate of the Inuit Tapirisat (Eskimo Brotherhood) of Canada.

must come first. I will not diminish anyone's right to be heard."

No one can question Mr. Justice Berger's efforts to ensure "the right to be heard." When, early on in the proceedings, officialdom in Yellowknife sent an order to Territorial civil servants not to testify before the Commission without the approval and instructions of the Assistant Commissioner of the Northwest Territories, Berger and his Inquiry staff forced the Territorial government to rescind the order. Similarly, they put pressure on the federal Department of Energy, Mines, and Resources and on DIAND to allow their employees to give evidence (and at government expense). A few weeks after the hearings opened, native and environmental groups complained to Berger that DIAND had curtailed their federal funds. There were conflicting reports and explanations on that one, and DIAND's current minister, Judd Buchanan, had to take a few hard knocks in the House of Commons. But, in the end, a compromise of sorts was worked out whereby native organizations got $100,000 and DIAND picked up enough of Berger Commission expenses to allow it to give $100,000 to an environmental group. Not surprisingly, CAGPL publicists at the Yellowknife hearings stated that the native peoples should not be allowed to present to the Commission their argument of "No land settlement, no pipeline," because the process of settling land claims could drag on for years, as had been the case in Alaska. But the Commissioner refused the suggestion, politely but firmly.

The principle that everyone had the right to be heard led to Berger's decision to take the Inquiry into various Mackenzie Valley and Yukon communities. It would be a tedious, tiring journey, but he felt it was the only way to get the whole story, particularly the human factors that lay buried beneath the interminable statistics and legal and technical jargon of the Explorer Hotel hearings. As he phrased it, "the native people who live on the land are the real experts." Since it wasn't possible for the people to come to Yellowknife and speak, he would go to their villages and fishing camps and listen. So, accompanied by a few aides, he turned up in community after community to hear what people had to say.

Usually dressed in a corduroy jacket with leather elbow patches, sportshirt, flannel trousers, and a scuffed pair of loafers, Mr. Justice Berger sat in Legion, community, and church halls - and sometimes at a table outdoors - explaining why he was there. "I want you, the people who live here, who make the north your home, to tell me what you would say to the Goverment of Canada if you could tell them what is in your minds. . . . The government is running the country. They can't come up here and listen to all of you. They can't conduct an inquiry themselves, so they sent me to do it. . . . I am going to make sure that when I present my report to the government, they understand what the consequences will be if a gas pipeline is built and a transportation corridor is established. . . . When you have the whole of the Mackenzie Valley and the northern Yukon, an area as large as Europe, people speaking six languages, people of four races, and environmental questions that have to be examined in depth if we're going to understand the consequences, it takes time. If Canada can't take time to make an informed decision on what's going to happen in our northland, then what's Canada got time for?"

Hearings started in the late afternoon when enough people had come together and the community leaders thought that it was time to start the proceedings. Meanwhile, the judge strolled around the village or camp, chatting with anyone willing to talk, or just taking in the scene. At the hearings, many a witness shuffled forward to where Berger was sitting, took the oath, and then, in a slow, self-conscious manner, told his or her story. Knowing that he was dealing with people who were basically shy and diffident, he always helped them to relax. "There's no rush, you just take your time," he assured the many who had never spoken in public before and were ill-at-ease and tense. Hearings lasted fourteen, fifteen, sixteen hours at a time and went on into the early hours of the morning, and some of the media people following the Commission's progress would yawn and stretch their arms and legs. But the judge stayed sitting, listening to everything being said to him, even when it was being spoken in Slavey, Dogrib, Chipewyan, Loucheaux, Inuktitut, and French and then had to be translated for him.

He stayed sitting and listening even when he was hearing the same things over and over again. He had spoken of everyone's right to be heard, and he would hear anyone with anything to say.

Each hearing lasted two or three days, until the judge and community leaders were satisfied that everyone had had their say. But he also spent time talking informally with people over a cup of coffee or at meal times. And the informality included baseball games played under the blistering light of the midnight sun at two and three o'clock in the morning between the home team and the visitors. (The latter, one of whom was Mr. Justice Berger, always lost.) In the cooler weather of the fall months, hearings were followed by drum dances, at which Berger attempted his own little chant-and-shuffle routine.

The serene manner in which Berger conducted himself during the initial community hearings quickly earned him the confidence of the native peoples. From Fort Smith, the old Territorial capital next the border of Alberta and the Northwest Territories to the Arctic-coast settlement of Paulatuk, from the community of Old Crow in the Yukon to Fort Franklin on Great Bear Lake, they poured out their hearts to him. Old and young said exactly what they felt, and it was an absolutely astounding outpouring. They spoke to him in story and song, in poetry and in prose. They offered him the wisdom of generations on the life of humans, plants, and animals. They explained to him the intricacies of ice, snow, permafrost. They told him about racism, alcholism, and sexual exploitation. They spoke to him with joy and warmth, with sadness and with anger. On the very first day of the very first community hearing, held in Aklavik, N.W.T., in April, 1975, he heard about the terrors a pipeline holds for the native peoples.

> Mr. Berger, I am Charlie Furlong, and I would like to speak to you today on behalf of my people. The people are not ready for land development. By people, I mean the Indians, the Eskimo and the Métis. The oil companies want to build a pipeline down the Northwest Territories. They want to take our gas, our oil. We will not even use our gas. It will go past our homes, our

communities, to heat southern Canada and our big brother, the United States of America.

The communities are not equipped to handle the men who will come. The health service is poor. Law enforcement is poor. Recreational facilities is poor, and education isn't 100%. Communities don't have a doctor. They have to go to bigger towns to get treated . . .

Kids in small towns go to a Grade 8 level, and then they have to go to Inuvik or Yellowknife to finish school. Most of them quit, or drop out, because they can't stay away from home. Native people themselves are not qualified for decent jobs during construction of the pipeline. White men will come from the south to build the pipeline, and again white men will come into the communities, take our daughters, our wives . . . .

I would like to see a land settlement between the Government and the people of the Northwest Territories, a land settlement where the native people will control their land and development. We are not against development, but we want to control it. In every movie about the Indian wars, the Indian people always lose. I now ask the Government, the southern people of Canada, to let us win this one.*

In Old Crow in the Yukon, Martha John Charlie, an old Loucheux woman, appealed to him,

God made this land, Mr. Berger, and now we who are old don't understand what white men are doing to it and who all the strangers are. We used to know all about the animals, the caribou and the muskrats, and now the animals act funny. Different, I think. This is why I'm against the pipeline. I'm thinking of our grandchildren. I want to know if the caribou will ever cross the pipeline, or if they'll go away and never come back. I want caribou for my grandchildren and great-grandchildren.†

At Fort Norman, N.W.T., Chief Paul Andrew described what it was like being a child at school in the 1950s and 1960s.

---

* Canada. Department of Indian and Northern Affairs, "Transcript of the Mackenzie Valley Pipeline Inquiry," Ottawa, 1976.

† McCullum, Hugh, "However, if we are forced to blow up the pipeline. . . ," Toronto, *The Canadian*, December 20th, 1975.

. . . I can still recall those days when it was brought up to us, entirely, continuously, over and over, that you got to speak English. Prior to those times, prior to coming to school, we had very little contact with any white people or people that understood the white language.

. . . one memory that always stands out when I think about those days that I started school was when this little kid did not know anything, any word in English, and he would get slapped around because he was using Slavey in school. That is the type of public school we had in Fort Norman, a foreign thing that we did not know anything about and yet we were expected to adapt to it, not in one year but in a few weeks . . . I don't know how I hacked it, but I managed to get up to seventh grade . . . therefore I had to be taken to an institution run by the Roman Catholic Church in Inuvik.

. . . The hostel life was not something that was enjoyable. Our way of life, our culture, our language, our form of identification, they were not there. Everything was southern-oriented. The meals, the language, sports, social life, any form of activities was all southern-oriented. And the proud way that we used to live, the proud way that we have existed without the white man's technology from time immemorial, that was being eliminated . . . the whole Territorial government programs were set up so that the native culture, the native identity, and the native language would be eliminated . . .†

At Fort McPherson, N.W.T., Phillip Blake summed up his five years of social work among the Dene.

First, I would like to say I am not an old man, but I have seen many changes in my life. Fifteen years ago, most of what you see as Fort McPherson did not exist. Take a look around the community now and you will start to get an idea of what has happened to the Indian people here over the past few years.

Look at the housing where transient government staff live. And look at the housing where the Indian people live. Look at how the school and hostel, the RCMP and Government staff houses are right in the centre of town, dividing the Indian people into two sides. Look at where the Bay store is, right on top of the highest point of land. Mr. Berger, do you think that that is the

† Canada. Department of Indian and Northern Affairs, "Transcript of the Mackenzie Valley Pipeline Inquiry," Ottawa, 1976.

way the Indian people chose to have this community? Do you think the people here had any voice in planning this community? Do you think they would have planned it so that it would divide them? Do you think they would have planned it so that it divided them and gave them a poorer standard than the transient whites who come in, supposedly to help them? . . .

Do you think that they chose to become cheap labour for oil companies, construction companies and government, instead of working for themselves and developing their own economy, in their own way?

In short, Mr. Berger, can you or anyone else really believe that we Indian people are now living the way we have chosen to live? Can you really believe that we have chosen to have high rates of alcoholism, murder, suicide and social breakdown? Do you think we have chosen to become beggars in our own homeland? . . .

For a while it seemed that we might escape the greed of the southern system. The north was seen as a frozen wasteland, not fit for the civilized ways of the white man. But that has been changing over the past few years. Now the system of genocide practiced on our Indian brothers in the south over the past few hundred years is being turned loose on us and our Eskimo brothers. "Don't be silly," you may say. "We are sorry about what we did in the past; we made some mistakes. But it's different now. Look, we give you an education, houses, and health services."

Mr. Berger, the system of genocide may have become a little more polished over the past few hundred years in order to suit the civilized tastes of the southern people who watch Lloyd Robertson on "The National." But the effect is exactly the same. We are being destroyed. Your nation is destroying our nation. What we are saying today, here and now, is exactly what Louis Riel was saying roughly a hundred years ago. We are a nation. We have our own land, our own ways, and our own civilization.

You and I both know what happened to Louis Riel. Yet now, a hundred years later, your prime minister is willing to say that Louis Riel was not all wrong. He is willing to say that, a hundred years later. But is he willing to change the approach that destroyed Louis Riel? And his nation? And is now threatening to destroy us? . . .

White people came as visitors to our land. Suddenly, they claim it as their land. They claim that we have no right to call it Indian land, land that we have occupied and used for thousands of years, which just recently the white man has come to visit. . . .

Is this the great system of justice, which your nation is so proud of? . . .

We have been satisfied to see our wealth as ourselves and the land we live with. It is our greatest wish to be able to pass this on, this land, to succeeding generations in the same condition that our fathers have given it to us. We did not try to improve the land and we did not try to destroy it. That is not our way. . . .

If your nation becomes so violent that it would tear up our land, destroy our society and our future, and occupy our homeland by trying to impose this pipeline against our will, then of course we will have no choice but to react with violence.

I hope we do not have to do that, for it is not the way we would choose. However, if we are forced to blow up the pipeline . . . I hope you will not only look on the violence of Indian action, but also on the violence of your own nation, which would force us to take such a course.

We will never initiate violence. But if your nation threatens by its own violent action to destroy our nation, you will have given us no choice.

Please do not force us into this position, for we would all lose too much.

Mr. Berger, I believe it is because I am a social worker here that I have . . . had to make some sense out of the frustration and desperation that people in this community and others along the valley are feeling. . . . It is clear to me that the pipeline in Alaska has not been any part of progress, whatever progress may mean. Where progress should mean people getting greater control over their own lives, greater freedom, the pipeline in Alaska appears to have driven people into the ground along with the pipeline. . . .

Mr. Berger, I guess what I am really trying to say is, can you help us? And can we help you make sure that the will of the people is respected? After all, isn't that what – isn't that supposed to be what Canada once stood for? Can we as an Indian nation help Canada to once again become a true democracy?*

Another Fort McPherson resident, Mrs. John Charlie, didn't hesitate to call a spade a spade.

To start off with, Mr. Berger, this land is ours. It's our land. We were on it for long before the white peoples, and when I think of

* *Ibid.*

this pipeline I just get sick of it. That's how much I don't want it to come through. Sure, we all know it's going to come through even though we don't want it to come through. It's going to spoil our land and its people, especially the young people. I say this because I see things with my own eyes through white people. Mr. Berger, some white peoples are real good and some are like dogs. I say this because I have girls. I see what's been done in Inuvik.

My sister is right here in this meeting. My niece was nice girl until one day one white man came along and told her that he was a single man. After she became an alcoholic, that white man was gone. One day when my sister was in Inuvik, I was with my sister in Inuvik. She had a phone call from Vancouver and someone asked her if she was Mrs. Greenland, and she said "Yes." This woman told her, "I heard your daughter is going out with my husband." And my sister told her, "I tried so hard to talk to my girls to be good, and you white women's husbands come down north and say they're single. They don't go only with my girls. They go with Eskimo girls, too. Why don't you white women come down with your husbands? Don't ever phone me like that again."

Mr. Berger, when the road came through to McPherson last year from Inuvik, the first taxi came to town. The next day two girls were missing. My cousin was looking for them, and here they were taken off with the taxi. All that makes me mad. That's why I don't like the pipeline. These are my nieces. They were just like my own girls. Now I worry about my own girls, how they will grow up. When I hear that there is going to be 800 people in every camp, I hope they make a law that the white people will have to stay away from the town. Like I said before, the white people are good, but some are no good.*

In Tuktoyaktuk, N.W.T., fisherman-hunter Vince Steen was sworn in and said,

First of all I'll introduce myself. I'm Vince Steen, brother to John Steen. What I have to say tonight, I don't represent anybody. I represent myself.

What I would like to point out, Mr. Berger, a lot of people seem to wonder why the Eskimo or "the people" don't take the white man's word at face value any more, why when he says he's going to drill an oil well and not spill any oil or make any mess, why

* *Ibid.*

they don't believe him any more. Well, from my point of view it goes way back, right back to when the Eskimo first seen the white man. Most of them were whalers, and the whaler wasn't very nice to the Eskimo. He just took all the whales they could get and never mind the results. Who is paying for it now? The Eskimo. There's a quota on how many whales he can kill now.

Then next, following the whalers, the white traders and the white trappers. The white traders took them for every cent they could get. You know the stories in every history book where they had a pile of fur as high as your gun. Those things were not fair. The native lived with it, damn well had to to get that gun, to make a life easier for himself.

Then there was the white trapper. He came along and he showed the Eskimo how to use the traps, steel-jawed traps, leg-hold traps. They used them. Well, they're still using them today, but for the first 70 years when they were being used there were no complaints down south about how cruel those traps are as long as there was white trappers using them. Now for the last five years they're even thinking of cutting us off, but they haven't showed us a new way to catch those foxes for their wives, though.

After them, after the white trappers and the fur traders, we have all the settlements, all the government people coming in and making settlements all over and telling people what to do, what is best for them. Live here. Live there. That place is no good for you. Right here is your school. So they did. They all moved into settlements and for the 1950s and 1960s they damn near starved. Most of them were on rations because they were not going out in the country any more. Their kids had to go to school. . . .

Then the oil companies. Well, the oil companies, I must say, of all of them so far that I've mentioned, seem to be the most – to have the most respect for the people and their ways. But it is too late. The people won't take a white man's word at face value any more because you fooled them too many times. You took everything they had and gave them nothing. You took all the fur, took all the whales, killed all the polar bears with aircraft and everything, and put a quota on top of that so we can't have polar bear when we feel like it any more. . . .

If they drill out there, if they finish off what little whales are left, what little polar bears are left, with one oil spill of any size big enough to hurt those animals, we're finished. The Eskimo population and culture is finished, because you have to live as a white man and you have nothing. You have no more seals to feed the foxes. You got no more fish to feed the seals, and you've got left no more seals to feed the polar bears, and the polar bears are going to go looking for some white men then, because they've got nothing left to eat. . . .

I would like to say in closing that the Eskimo is asking for a land settlement because he doesn't trust the white man any more to handle the land that he owns and he figures he's owned for years and years. . . . I cannot see where a white man or any government can turn it down, seeing we're not asking to claim the land for ourself. We're asking to share it, but share it on a 50-50 basis, not on a 100% basis like it's been going for the last 50 or 60 or 70 years. Thank you.*

In Whitehorse, the capital of the Yukon, taxi-driver Rainer Genelli was fed up with the whole business of the North being a milch cow for the rest of Canada.

". . . what really bothers me is this sort of whole imposition of southern values on the north, imposition of like, you know, people in the north having to face up, "Well, lookit, it's progress, you can't stop it" or something. Well, I just sort of always think, well, I wonder how people in Toronto would react if the people of Old Crow went down to Toronto and said, "Well, look, we are going to knock down all of these skyscrapers and highrises, and, you know, blast a few holes for lakes to make for muskrat trapping, and you people are just going to have to move out and stop driving cars and move into cabins." I mean . . .

. . ."And you told them that was progress," prompted Judge Berger.

"Yes, right," agreed Genelli, "and told them that it was progress. Well, muskrat skins are $5,000 apiece and, you know, Toronto is a beautiful muskrat breeding ground or something, or else maybe an important inquiry for establishing a trapline in the Parliament buildings or something."†

At Inuvik, N.W.T., and at Yellowknife, N.W.T., Dr. Peter Usher, a social scientist, offered a personal and a professional view of northern versus southern life styles.

---

* *Ibid.*

† *Ibid.*

I remember a time when to come to this north country was to come to a really different place, and nobody had to tell you the difference. You knew it. It wasn't possible then to live as a parka-clad suburbanite. It wasn't possible to pick up the phone and call the folks back home. It wasn't possible to visit four settlements a day on some kind of junket. When you went some place you stayed a while, because you couldn't just pick up and go. That's the way life was.

. . . I think the biggest change I've noticed is that southerners can now come north and live in a place like Inuvik and maybe even Tuk [toyaktuk] and Aklavik and they can insulate themselves from the real north. They don't ever have to travel on the land. They don't ever have to eat local food. They need never be without running water and electric power, and they need never even meet local people except in a very formal way, you know, by their business or by their work.

. . . the society and the economy that southern Canadians have, by and large, chosen to build here is for their own peculiar purposes, deliberately insulated from the land and pre-existing life of the north, with little reference to native people of this land and less knowledge of them. . . .

I am also sick and tired of the Federal and Territorial government mythology about the three founding races all working side by side to build a new north. Did the Indians and Eskimos ask all those others to come here and help them do that? Will the new north really be built in a co-operative and equitable way? Will the Indians and Eskimos who are the majority here really control the pace and direction of northern development in the evolution of northern society? Sir, I think a stroll around this town will give you the answers right away. The white man made the plans for this place, and he makes the rules. . . . White people run the businesses here. They run the Town Council. They set the priorities. They determine the future. Now, what kind of co-operation is that? . . . if there is one pervasive anxiety among native residents of the smaller communities around here it is that their villages may one day become like Inuvik. But unless the whole process of northern development changes, and fast, Inuvik is indeed the way of the future. . . .

By the 1950s the fur-trade economy was in collapse. The response of the rest of Canada was to shore up the native economy with wages, welfare, and education. . . . it's easy for the casual observer to assume that the old ways are dead and, with them, native involvement with the land. Indeed, that assumption has provided the basis for government policy in the north for at least 25 years. . . .

Despite almost complete urbanization, despite a massive government drive to provide alternate sources of income, chiefly through wage employment . . . some communities are harvesting more fur than ever. Throughout the region, country food is common by far at mealtimes. Those who have steady jobs are often to be found out on the land on their weekends and holidays. . . . People like country food better than store-bought food. . . . In their opinion, country food not only tastes better but it is also more satisfying and nutritious. . . . The north may well be the only place where a poor man's table is laden with meat as a matter of course.

. . . there is also a deep rooted social and cultural reliance on the land. To native people, the land is more than just a source of food or cash. It is the permanent source of their security and of their sense of well-being. It is the basis of what they are as people. They know that the land and the birds, fish and animals it supports have sustained them and their ancestors since time immemorial. Properly cared for, they feel it can always do so. . . . Native peoples' concepts of freedom and self-determination are closely linked with the integrity of the land. "Free life" is a commonly heard expression with regard to life on the land. Those who do not now live on the land, who work and live in the settlements, still regard the land as an essential part of their lives. Few fail to express an awareness that the security of town life was bought only at the cost of this "free life." The weekend or holiday trapper and hunter is a growing phenomenon in the Western Arctic, as wage employment becomes more widespread. The importance of this phenomenon is vastly underestimated by outsiders who see it as a mere pastime or sport, or who see it as inefficient or unspecialized. It is in my view absolutely essential to the well-being of Western Arctic society. . . . it is necessary for the social well-being of individuals.

There have been many enormous changes wrought in the Western Arctic in this century. Until recently, all of these were restricted to town life; in geographical terms to a few tiny dots on the enormous map of the north country, a few square miles out of hundreds of thousands. The land itself remained untouched. The rules of town life did not apply on the land. The land was a refuge. The weekend hunters go out partly because they need the food. They also go to get back to an older, simpler, but more interesting and rewarding way of life. On the land, men who work all week for someone else suddenly become their own bosses again. They can work according to their own rules and disciplines, not the half-understood ones imposed by the white man's wage system. On the land, men can operate by the time they understand, the time dictated by nature, not by the clock. On the land, the humdrum and the pressure of town life disappear. . . .

In my experience, the attitudes of large numbers of native people to wage employment have not changed significantly in recent years. Rarely do they have a lifetime commitment to wage employment, or to the associated attitudes toward saving, spending, and career planning implicit in this way of life. The economic history of the Western Arctic has been one of boom and bust from the whaling and fur-trade days to the present. Little or nothing in the experience of most native people has given them reason to share the economic assumptions of most other Canadians. The ideas of moving inevitably up a career ladder, of having one's pay cheque increase by so many percent per year, of getting a new couch this year and a color T.V. the next . . . are not current among native northerners.

Jobs, then, are a temporary resource to be exploited towards specific ends. Many native northerners seek and then leave jobs once they have earned enough for some specific purpose. Jobs are not valued for their own sake, but as temporary strategies. Given that the white man's economy appears to many native people as an erratic and unreliable phenomenon, the land and its resources are by comparison permanent. This attitude is prevalent even among those native people who are permanently employed and who prefer wage employment and town life to a more complete dependence on the land. They are by no means convinced that they will have – or want – steady wage work forever, and hence see the land as insurance against the day when employment may not be available. In their view, the white man's economy has its ups and downs, jobs come and go, but the land is always there.

It is, as many have said to this Inquiry, their "bank." Apparently they do not want this bank or their ability to use it disrupted. I believe this is why many native people are of the view that full-time but short-term employment, since it disrupts the routine and capitalization of hunting, has in the longer term

> a detrimental effect on their well being.*

But it was the proud, bitter words spoken to the Commission by Frank T'Seleie at Fort Good Hope that did more than call a spade a spade. It was more like calling a spade a bloody bulldozer.

> . . . This is the first time in the history of my people that an important person from your nation has come to listen and learn from us, and not just come to tell us what we should do, or trick us into saying "yes" to something that, in the end, is not good for us. I believe you are an honest man. I believe you are a just man, Mr. Berger, and that you do not intend to be a part of a plot to trick us or fool us or play games with us. . . .
>
> Mr. Berger, you have visited many of the Dene communities. The Dene people of Hay River told you that they do not want the pipeline because, with the present development of Hay River, they have already been shoved aside. The Dene people of Fort Franklin told you that they do not want the pipeline because they love their land and their life and do not want it destroyed. Chief Paul Andrew and his people in Fort Norman told you that no man, Dene or white, would jeopardize his own future and the future of his children. Yet you are asking him to do just that if you asked him to agree to a pipeline through this land. Phillip Blake, in Fort McPherson, told you that if your nation becomes so violent as to force a pipeline through our land, then we love our land and our future enough to blow up the pipeline. He told you that we, the last free Indian nation, are willing to fight so that we may survive as a free nation.
>
> You have heard old people and young people, Mr. Berger. You have heard people who were raised in the bush and people who were raised in government hostels. You have heard men and women, people who have worked for the white man, and people who have never sold their labour. People from the Mackenzie Delta to the Great Slave Lake. People have talked to you from their heart and soul, for they know, as I know, that if a pipeline goes through they will be destroyed.
>
> All these people have told you one thing, Mr. Berger. They have told you that they do not want a pipeline. My people are very strong, Mr. Berger, and we are becoming even stronger. My

* *Ibid.*

people are finding new strength for the struggle that we are going through. That is why I can say to you, Mr. Berger, "Yes, we can stop the pipeline." Our grandchildren will remember us, the Dene people here today, and the Dene people who have talked to you in other communities, as the people who stopped the pipeline from coming through their land. . . .

We do not want to have to fight and struggle forever, just to survive as a people. Your nation has the power to destroy us all tomorrow if it chooses to. It has chosen instead to torture us slowly, to take our children from us and teach them foreign ways and tell us that you are teaching them to be civilized. Sometimes now, we hardly know our own children. You have forced us into communities and tried to make us forget how to live off the land, so you could go ahead and take the resources where we trap and hunt and fish. You encourage us to drink liquor until we are half crazy and fight among ourselves. What else other than liquor is the Territorial Government willing to subsidize to make sure that prices are the same throughout the Northwest Territories? Does it subsidize fresh food or clothing or even pop in the same way? No. Only liquor. Try to buy anything else at Yellowknife prices throughout the North. The government knows very well that liquor helps keep my people asleep, helps keep them from realizing what is really happening to them and why. I know very well too, Mr. Berger, because I used to drink. . . .

And Chief T'Seleie didn't hesitate to come down like a ton of bricks on Robert Blair, the president of Foothills Pipe Lines Ltd. of Calgary, who happened to be present at this community hearing.

You are like the Pentagon, Mr. Blair, planning the slaughter of innocent Vietnamese. Don't tell me you are not responsible for the destruction of my nation. You are directly responsible. You are the twentieth-century General Custer. You have come to destroy the Dene nation. You are coming with your troops to slaughter us and steal land that is rightfully ours.

You are coming to destroy a people that have a history of thirty thousand years. Why? For twenty years of gas? Are you really that insane? The original General Custer was exactly that insane. You still have a chance to learn, a chance to be remembered by history as something other than a fool bent on destroying everything he touched. You still have a chance. You have a choice. Are you a strong enough man to really exercise

> your freedom and make that choice? You can destroy my nation, Mr. Blair, or you could be a great help to give us our freedom. Which choice do you make, Mr. Blair? Which choice do you make for your children and mine?*

Chief T'Seleie may have identified the wrong man as General Custer. When both Foothills Pipe Lines Limited and Canadian Arctic Gas Pipeline Limited applied to the settlement council of Fort Good Hope for land-use permits to conduct pipeline-survey work, the council refused to issue the permits but had its decision overruled by Mr. Stuart Hodgson, the Commissioner of the Northwest Territories. (He happens to be an appointed, not an elected official, but had been quoted as saying, "I am the Government.") It is to Mr. Blair's credit that, in this situation, he cancelled all Foothills' survey work in and around Fort Good Hope.

At Rae, on the northern arm of Great Slave Lake, Georges Erasmus, president of the Indian Brotherhood of the N.W.T., explained to Berger the whys and wherefores of the Dene movement for independence:

> Our struggle is for self-determination. We want to be in charge of our lives and our future. Very recently, the Prime Minister, in addressing the Queen, stated that any time there is an element within Canada that struggles to preserve its integrity, Canada as a whole is being protected.
>
> I think we agree that our struggle is in the interests of Canada as a whole. We are willing to settle for nothing less than the right to direct our lives. Last year at Fort Simpson, at the joint general assembly of the Métis Association and the Indian Brotherhood, we passed the Dene Declaration. We stated we were a nation. Some people thought this was a new position. But there is nothing new in that idea. We were simply stating the same position that our people have always had. It was the same position that the leaders at the signing of Treaties 8 and 11 had.
>
> I'm not talking about the version that's written in the treaties. What's written on the pieces of paper that represent what is recorded in Canadian history as Treaties 8 and 11 is that the Dene of the valley sold their land, gave up their right to govern themselves. That is not the kind of treaty our people passed.

* *Ibid.*

Our people have never given up the right to govern themselves. Our people have never given up this land. The reason that the native organizations were formed, the reason that the Indian Brotherhood was formed, was for the same thing. Our people were experiencing that our version of the treaty was not being met even though we had never given up the land, even though we had never given the right for somebody else to make decisions for us. . . . We needed the organizations to regain that kind of authority. The caveat hearings with Judge Morrow, the testimony there, the evidence there, it's consistent. Our people did not give up the land, did not give up the right to govern themselves. . . .

Our position is, there can be no pipeline until after our land claims [are settled]. Again, the reason why I started out by saying that this Inquiry is not our last hope is because, if we are going to survive – and we are going to guarantee that in a hundred or two hundred years there are going to be Dene – that can only be guaranteed by our actions, the actions of the Dene to make that happen.

There is no question, there can possibly be no question on whether or not the Dene nation exists. And I think that Mr. Berger, you, probably more than anybody else, know that there exists a Dene nation. There has existed a Dene nation for a long time, and this can happen to be the case for a long time in the future if we, the Dene, decide we want to remain Dene. And that is what we intend to do.*

One of the last community hearings was held at Lac la Martre, about a hundred and forty miles northwest of Yellowknife. Next to this small stretch of water live a couple of hundred Dogrib. Late on the warm evening of August 13, 1976 in the settlement hall, the last witness had finished testifying and the Commissioner had adjourned the hearing. He was preparing to leave when a young man walked up and asked if he could possibly add his two cents' worth. Some thoughts he had written down the night before were bothering him, and he had to get them off his chest. After taking the oath on the bible held out to him by the secretary to the Commission, Jim Green, the settlement manager, had his say. It was short and simple, and drew muted sounds of approval and agreement from older members of the audience

* *Ibid.*

when translated into Dogrib. In a way, it summarized much that had been said to the judge in the course of the whole Inquiry.

> I am obviously not Dene. I don't have a Dene language, a Dene mind, or a Dene education. . . .
>
> I am a white man. I am a transplanted European white man, but I was born in this country. My father was born in this country. Canada is my home; no place else. This is my home. . . .
>
> I think one of the reasons that I've been able to talk is because I've heard so many people tonight talking that I've never heard talk before, and talking from their hearts, and talking what they really feel about something that is very important. . . .
>
> This pipeline we are talking about is the latest example, I think, of a long series of mistakes that new people in this country have made, because they are thinking more about money than they are about people. I think this pipeline is another example of a mistake that could be made, that could destroy what little is left of the country that I call my country . . .
>
> I guess the whole thing began a long time ago with our ancestors coming to this country. They took the cream off the European continent. Then they came to this country, and they took the top off of this country, and they made money as fast as they could . . . and it just went on and on and on until today. And look what's left in North America. Not very much.
>
> And all the time we were right. We were right the whole way. God was on our side. We were right. The Indians didn't have the technology to develop the land, to develop the resources, so we did it. And we put all the profits in the bank.
>
> Looking back on it, as far as I can see, the white North Americans' reaction to this country was fear. They were afraid of starving. They were afraid of the savage natives that were roaming in the country, and they were afraid of the unknown wilderness. And it looks like this same reaction is carried on right to today. We still want to control the world, the weather, the earth, all the living things on it. It seems like we see it as a triumph to conquer the wilderness and control everything that lives in it.
>
> And I know this isn't a very popular thing to say, but I think we've been wrong. Right from the start I think we've been wrong. I think we should have been learning to live with things, like in harmony with everything else. We should have been developing a lifestyle that would complement everything else,

that would complement all other living people, and all other living things on the earth. And I think in that way we could have had some hope of survival. But I think we blew it. . . .

Look what we've done on the North American continent in two hundred years. . . . we fenced it in; we grazed it; we logged it off; we dug holes in it; we blew holes in it; we levelled it off; we cemented it over. And it got to the point where we had to create National Parks . . . so that we had something left to look at. Something to remind us of how it used to be, or maybe how it could have been.

And all this time we had so much trouble with the Indians. They didn't give up. They kept coming back. The vanishing race we called them. The vanishing race that kept not vanishing. How we tried. We declared them savages, heathens, non-citizens, wards of the government. We pushed them aside, and we developed the hell out of this country.

So, now, it looks like we have to finish the job. It looks like there's to be no end to this madness. There'll be no end until we die of lack of food, lack of water, lack of air, under garbage heaps that we make ourselves.

. . . I began by saying that I am a white man, and so I had to speak about the pipeline as a white man. And then I went on to support the Dene people, who have been speaking to you about the pipeline. They have been speaking to you about the pipeline for months.

And I found that I had to say this tonight because . . . what I've been hearing they've been saying, and what I've been reading they've been saying to you for months, it just makes sense. It just makes so much sense that I just can't see it any other way. They're thinking about their children. And I think about my children. And when I hear what they're saying about – what they think about – what they're afraid about the future of their children, I'm afraid of the same things. . . .

Well, I am thinking about my children when I ask you, and the government, and the people of this country to stop and think awhile. Maybe we don't need this pipeline next year. Maybe we don't need it in five years. Possibly we don't need it at all. And it's what these people have been saying all along, and I just have to agree with them. Maybe we should just stop while there is something left.*

* * *

---

* *Ibid.*

In Yellowknife in November, 1976, after listening to the 1717th and last witness, Mr. Justice Thomas Berger adjourned the final hearing of his Inquiry. As he did so, he casually remarked to those present, "You will be hearing from me." Assembling his staff after lunch that same day, the judge backed up his statement by placing on a table for their inspection his report in miniature, written in longhand. Barely six months later, an expanded version of it was tabled in the House of Commons and then released for public examination in the form of a large, handsomely illustrated book.

Into the 248 pages of *Northern Frontier, Northern Homeland: The Report of the Mackenzie Valley Pipeline Inquiry, Volume One*,* the Commissioner compressed much of the evidence of 281 volumes of testimony, some 40,000 pages in all. The author is on record as saying that all along he has been "guided by the conviction that this Inquiry must be fair and it must be complete." Well, *Northern Frontier, Northern Homeland* is a clear, concise summary of various aspects of pipeline development, history, culture, wildlife, education, economics, and engineering. The book is, in fact, a crash course in northern studies. "Complete" it certainly is, but "fair"? Oil companies who have spent millions of dollars on exploration and environmental studies won't think so. The officials and businessmen who run the Northwest Territories won't think so. Energy-hungry Canadians and Americans won't think so. Because Mr. Justice Berger has recommended there be no pipeline for at least ten years and that the "last frontier" in North America be handed back to its native peoples.

The Commissioner of the Mackenzie Valley Pipeline Inquiry is a forthright man. At the very beginning of the *Report*, he states,

> We are now at our last frontier. It is a frontier that all of us have

---

* Volume 1 dealt with the broad social, economic, and environmental impacts that a gas pipeline and an energy corridor would have in the Mackenzie Valley and the Western Arctic. Volume 2, published a few months later, set out the terms and conditions that, Mr. Justice Berger recommended, be imposed by the federal government should a pipeline be built.

read about, but few of us have seen. Profound issues, touching our deepest concerns as a nation, await us there.

The North is a frontier, but it is a homeland, too, the homeland of the Dene, Inuit and Métis, as it is also the home of the white people who live there. And it is a heritage, a unique environment that we are called upon to preserve for all Canadians. ...

The North is a region of conflicting goals, preferences and aspirations. The conflict focuses on the pipeline. The pipeline represents the advance of the industrial system to the Arctic. The impact of the industrial system upon the native people has been the special concern of the Inquiry, for one thing is certain: the impact of a pipeline will bear especially upon the native people. That is why I have been concerned that the native people should have an opportunity to speak to the Inquiry in their own villages, in their own languages, and in their own way. ...

The issues we face are profound ones, going beyond the ideological conflicts that have occupied the world for so long, conflicts over who should run the industrial machine and who should reap the benefits. Now we are being asked: How much energy does it take to run the industrial machine? Where must the energy come from? Where is the machine going? And what happens to the people who live in the path of the machine?

... In the days of the fur trade, the native people were essential. In the north today, the native people are not essential to the oil and gas industry, and they know it. The outside world may need the North's oil and gas resources, but it does not need the native people to obtain those resources. Outsiders know exactly what they want and exactly how to get it, and they need no local help.

Mr. Justice Berger is also a reasonable man. He knows perfectly well that it's no fun freezing to death – in the dark. Life is a matter of "live and let live." So he observes that,

I have proceeded on the assumption that, in due course, the industrial system will require the gas and oil of the Western Arctic, and that they will have to be transported along the Mackenzie Valley to markets in the South. I have also proceeded on the assumption that we intend to protect and preserve Canada's northern environment, and that, above all else, we intend to honour the legitimate claims and aspirations of the native people.

What, then, are his objections to a "go" decision on a gas pipeline before 1987? There's no denying that he is more people-oriented then he is power-oriented. His opening statements are proof enough of that. But he is also the jurist, the seeker after that which is right, just, and proper, and certain key phrases and sentences in *Northern Frontier, Northern Homeland* jump off the pages to substantiate this. He hits the nail right on the head when he notes "The risk is in Canada. The urgency is in the United States." He points out that "the proposed natural gas pipeline is not to be considered in isolation . . . if a gas pipeline is built, an oil pipeline will follow . . . we must consider, then, the impact of a transportation corridor for two energy systems, a corridor that may eventually include roads and other transportation systems. . . . The cumulative impact of all these developments will bring immense and irreversible changes to the Mackenzie Valley and the Western Arctic." He warns that a gas pipeline is "not just a 120-foot right-of-way. It will be a major construction project across our northern territories."

There is, too, the complication of the northern Yukon, an arctic and subarctic wilderness of incredible beauty and of equally incredible fragility. Here are the calving grounds of the Porcupine herd, one of the last great caribou herds in North America. Here in the northern Yukon, hundreds of thousands of migratory wildfowl find food and shelter each summer and fall. Here are mountains and valleys, spruce forests and lakes, tundra and arctic seacoasts, a "unique ecosystem" spread over nine million acres that has survived simply because of the inaccessibility – and hence the tranquility – of the region. Yet the Arctic Gas pipeline proposal to carry gas from Prudhoe Bay, Alaska, would cross this region, either along a coastal or an interior route. And what's to stop a gas line being followed by an energy corridor – an oil pipeline, a highway, and other developments? "If this unique area of wilderness and its wildlife are to be protected, the Arctic Gas pipeline should not be built across the northern Yukon. The region should not be open to any other future proposal . . . I therefore urge the Government of Canada to reserve the northern Yukon as a wilderness park." If there has to be a Canadian right-of-way for the transport of American gas, the Commissioner thinks a route through the southern Yukon might be feasible – provided social and economic impacts are assessed and native claims are settled.*

On wildlife and human grounds, he advises against an energy corridor across the mouth of the Mackenzie Delta and is extremely wary of the continuance of offshore oil-and-gas exploration work until there are successful techniques in controlling or cleaning up a major spill in conditions of floating ice or rough water. "Canada has chosen to pioneer offshore oil and gas exploration in the Arctic. . . . Canadians have a grave responsibility in this matter." Bird and whale sanctuaries are needed, the latter inviolate to petroleum exploration at any time. Like the northern Yukon, the Mackenzie Delta and Beaufort Sea region supports a "unique and vulnerable arctic ecosystem. Its wildlife has been a mainstay of the native people of the region for a long time, and still is today."†

From a strictly environmental point of view, Mr. Berger does believe it feasible to build a Mackenzie Valley pipeline. He notes that no major wildlife populations would be threatened, and no wilderness areas would be violated, although sanctuaries would be needed for migratory wildfowl and already endangered falcons. But two major factors force

---

* On April 19, 1977 – just prior to the public release of the Berger Report – the Canadian government hastily ordered an inquiry into the Alcan Pipeline Company proposal to bring Alaskan gas in a line alongside the Alaska Highway through the Yukon, British Columbia, and Alberta to U.S. markets. According to the backers of the Alcan project, if this line is built, Delta-Beaufort Sea gas can be pipelined across the Yukon at a later date and fed into the Alcan pipeline. Kenneth Lysyk, dean of law at the University of British Columbia, was appointed chairman of the board of inquiry, which was required to report by August, 1977 on the socio-economic impact of the Alcan line. A separate group of inquiries, appointed by the Minister of the Environment, examined the environmental impact. Thus, Dean Lysyk and the ministerial appointees had three and a half months to do what it took Berger and staff two and a half years to evaluate.

† In May, 1977, the Trudeau government approved the resumption of drilling for oil and gas in the Beaufort Sea by a subsidiary of Dome Petroleum Ltd. of Calgary, Alta.

him to recommend a ten-year delay in pipeline construction in the Valley.

First, there are "critical gaps in the information available about the northern environment, about environmental impact, and about engineering design and construction on permafrost terrain and under arctic conditions. The question of frost heave is basic to the engineering design of the gas pipeline. Both Arctic Gas and Foothills propose to bury their pipe throughout its length, and to refrigerate the gas to avoid the engineering and environmental problems resulting from thawing permafrost. But where unfrozen ground is encountered, in the zone of discontinous permafrost or at river crossings, the chilled gas will freeze the ground around the pipe, and may produce frost heave and potential damage to the pipe. The pipeline companies are obviously having trouble in designing their proposal to deal with frost heave. . . . It is likely that the companies will make yet further changes in their proposals, changes that are likely to increase costs and to alter substantially the environmental impact of the project." The second is the culture, values, and traditions of the native peoples, which "amounts to a great deal more than crafts and carvings." Their institutions, values, and language have been "rejected, ignored, or misunderstood." And while the native and non-native populations in the Western Arctic are about equal in number, in Berger's opinion, it is the native people who are the permanent population. "There they were born, and there they will die." So the future of the North should not be determined only by southern ideas of frontier development but also by the ideas of the people who call it their homeland. And he found clear evidence in the course of the Inquiry that the more the industrial frontier displaces the homeland in the North, the greater the incidence of social pathology. "Superimposed on problems that already exist in the Mackenzie Valley and the Western Arctic, the social consequences of the pipeline will not only be serious – they will be devastating." Berger's great concern is the social cost of building a pipeline. He is quite convinced that the advance of the industrial system to the frontier "will not be orderly and beneficial, but sudden, massive and overwhelming."

The only way to avoid this chaos is, Berger says, settlement of land claims before a pipeline is built. That is what the peoples told him time and time again. They want to determine their own place in, not assimiliation into, Canadian life. They want to *begin* with a land settlement. From there, they'll go on to decide other matters: renewable and non-renewable resources; education; health and social services; and, above all else, a political re-shaping of the North. They are not, Berger emphasizes, renouncing Canada or Confederation. But they do want their children and their children's children to be secure in the knowledge of who they are and where they came from. "The idea of new institutions that give meaning to native self-determination should not frighten us," Berger reports. "In the past, special status has meant Indian reserves. Now the native people wish to substitute self-determination for enforced dependency."

* * *

The Honourable Mr. Justice T. R. Berger is convinced that, if the Trudeau government accepts his recommendations, a Mackenzie Valley pipeline can be built "at a time of our choosing, along a route of our own choice." He feels sure that, with the passage of time, it may well be possible to reconcile the present claims of the native people with the future gas-and-oil requirements of other Canadians. Well now, there's going to be hot and heavy argument about *that* in northern Canada, in southern Canada, and in a frightened, fuel-hungry United States. Mr. Justice Berger is going to be hailed as a hero and damned as a do-gooder. His basic recommendation of a ten-year delay – plus the various other opinions he has expressed in his *Report* – undoubtedly will be contested, scorned, and repudiated. But something touched on in the *Report* cannot be contested, or scorned, or repudiated by Canadians. Berger emphasized it in a speech in 1975 to students at Queen's University, Kingston, Ontario.

> . . . It is natural for us to think of developing the North, of subduing the land, populating it with people from southern Canada, and extracting its resources to fuel Canada's industry and heat our homes. Our whole inclination is to think in terms of

expanding our industrial machine to the limit of our country's frontiers.

But the native people are saying to us: Why do you say the north is your last frontier? Why should you develop it? . . . They say: We have lived here for thousands of years. We are the majority. What right have you to tell us what the future must hold for us? What right have you to exploit the resources of the land where we live? It is a question being asked of the white race all over the world. And it is being asked of us by northern native peoples here in our own country.

. . . maybe it is time the metropolis listened to the voices on the frontier, time the metropolis realized it has something to learn from Old Crow and Hay River. Because what happens in the north will be of great importance to the future of our country. It will tell us what kind of people we are.

Acknowledgements

Mrs. Carol Waldock typed drafts of each chapter more times than either of us probably cares to remember. However, before she could do this, she had to decipher my scribblings and scrawlings, which often wandered onto the back of every second foolscap sheet of paper. I am deeply grateful to Mrs. Waldock for her transliteration and typing skills. And I consider it an unexpected bonus that she had stamina enough left to run a last-minute copyediting eye over the final version of the manuscript.

There is an enormous body of literature on the fur trade, and I have benefited from the research and writings of many persons. But I am particularly indebted to the publications of Professor Harold A. Innis and Dr. W. Stewart Wallace of the University of Toronto, Dr. W. Kaye Lamb, former Dominion Archivist, Mr. R. M. Patterson of Victoria, British Columbia, Professor E. E. Rich of St. Catharine's College, Cambridge University, England, Professor John S. Galbraith of Stanford University, California, and to those of that remarkable American man of letters, Mr. Bernard De Voto. This book owes much to their investigations of the trade and its practitioners.

I also want to record my gratitude for the wealth of information I discovered in Mr. Gary D. Sealey's unpublished M.A. thesis, "A History of the Hudson's Bay Company, 1870-1900" (University of Western Ontario, 1969) and in Professor Maurice Zaslow's *The Opening of the Canadian North, 1870-1914* (McClelland and Stewart Limited, Toronto, 1971).

A remarkable number of people have written accounts of the Mackenzie River region. All of them have enlightened my ignorance. However, for my purposes, a few writers were singularly perceptive and informative. Miss Agnes Deans Cameron sketched memorable portraits of the river's peoples in *The New North* (D. Appleton and Company, New

York and London, 1909). In *Sixty Below* (Jonathan Cape, Toronto, 1944), Sergeant Anthony Onraet of the Canadian Army gave me a definite "feel" for the Peace, Slave, and Mackenzie rivers. Father René Fumoleau's *As Long As This Land Shall Last* (McClelland and Stewart, Toronto, 1973) documents the fumbling, bumbling, way in which the government of Canada has generally treated its Indian and Métis wards. And Mr. Martin O'Malley's *The Past and Future Land* (Peter Martin Associates, Toronto, 1976) made it crystal clear that, in its community hearings, the Berger Inquiry was a humbling experience for any Canadian not of Dene, Métis, or Inuit ancestry.

I received various kinds of help from various individuals. For this, I offer thanks to Miss, Mr., or Ms. Jackie Addersley, Clerk, Corporation of the Town of Fort Smith, N.W.T.; Mr. Art Bevington, the airport manager at Fort Smith, N.W.T.; Ms. Maureen G. Bishop, Travel Counsellor, Travel Alberta, Edmonton, Alta.; Mr. John A. Bovey, Provincial Archivist, Province of Manitoba; Mrs. M. E. Braathen, Travel Counsellor, TravelArctic, Division of Tourism, Government of the Northwest Territories, Yellowknife, N.W.T.; Ms. Robin Chambers, TravelArctic, Division of Tourism, Government of the Northwest Territories, Yellowknife, N.W.T.; Mr. Roger Comeau, Head, Pre-Confederation Section, Manuscript Division, Public Archives, Ottawa, Ont.; Mrs. Darlene J. Comfort, Fort McMurray, Alta.; Mr. Henry G. Cook, Coordinator of Historical Programs, Government of the Northwest Territories, Yellowknife, N.W.T.; Father Ebner, Director of the Northern Life Museum, Fort Smith, N.W.T.; Mr. J. D. Fecht, Settlement Manager, Fort Providence, N.W.T.; Mr. R. H. Fenner, Imperial Oil Limited, Toronto, Ont.; Mayor R. M. Findlay, The City of Yellowknife, N.W.T.; Sister Mary Josephine Fox and Father René Fumoleau of St. Patrick's Church, Yellowknife, N.W.T.; Mr. D. Kieler, Head, Territorial Parks, Business Services and Tourism Division, Department of Economic Development and Tourism, Government of the Northwest Territories, Yellowknife, N.W.T.; Mr. Frank Lewis, Forest Officer II, Fort McKay, Alta.; Mrs. Nancy Lonnay of the Pipeline Co-ordination Division, Department of Indian and Northern Affairs, Ottawa, Ont.; Mr.

J. Mackay, Imperial Oil Limited, Toronto, Ont.; Mrs. N. J. Mackie, Secretary-Treasurer, The Town of Hay River, N.W.T.; The Manager, Hydrographic Chart Distribution Office, Marine Sciences Directorate, Department of the Environment, Ottawa, Ont.; Miss Gillian Mars, Library Technician, Hudson's Bay Company, Hudson's Bay House, Winnipeg, Man.; Mr. John S. L. McEwen, Settlement Manager, Fort Norman, N.W.T.; Miss Penny McLaren, Pre-Confederation Records Section, Manuscript Division, Public Archives, Ottawa, Ont.; Mr. Robert V. Oleson, Public Relations Officer, Hudson's Bay Company, Hudson's Bay House, Winnipeg, Man.; Mr. D. P. Ormond, Clerk-Administrator, Town of Fort St. John, B.C.; Mrs. Joan Parker of the Gage Library; Mr. Stewart Robinson, Traffic Manager, Northern Transportation Company Limited, Edmonton, Alta.; Ms. Merle Rudiak, Syncrude Canada Ltd., Fort McMurray, Alta.; Mr. L. A. Wilderspin, Settlement Manager, Norman Wells, N.W.T.; and Mr. Tudor Williams, Co-ordinator, Educational Services, Syncrude Canada Ltd., Edmonton, Alta.

The following publishers, individuals, and institutions have kindly allowed me to quote from their publications or public releases. (Bibliographical information on the publications themselves is given in relevant footnotes within the book.)

The Anglican Book Centre, Toronto, Ont.
The Honourable Mr. Justice T. R. Berger, Vancouver, B.C.
Dr. John W. Chalmers, Edmonton, Alta.
The Champlain Society, Toronto, Ont.
Hurtig Publishers, Edmonton, Alta.
Imperial Oil Limited, Toronto, Ont.
The Department of Indian Affairs and Northern Development, Ottawa, Ont.
Peter Martin Associates Limited, Toronto, Ont.
Mr. J. C. MacGregor, Edmonton, Alta.
McClelland and Stewart Limited, Toronto, Ont.
Mr. Hugh McCullum, Toronto, Ont.
Press Porcépic Ltd., Erin, Ont.
Mr. Wade Rowland, Toronto, Ont.

J.K.S.

Further Reading

(In this list of books, an asterisk indicates a paperback edition.)

ALEXANDER MACKENZIE
Just about every biography of Mackenzie concentrates on little else but his explorations. And just about all his biographers have idolized him. To these writers, he was Superman: daring, dauntless, and unblemished by normal human faults and failings. The same is true of those who have produced the many editions of the *Voyages from Montreal* — with one marvellous exception. W. Kaye Lamb's edition of *The Journals and Letters of Sir Alexander Mackenzie*, published in 1971 for the Hakluyt Society by the Cambridge University Press and marketed in Canada by Macmillan of Canada, Toronto, contains, in its Introduction, an excellent analysis of Mackenzie's character, life, and work.

DAVID THOMPSON
The best source of information on David Thompson happens to be David Thompson. The reader is advised to consult *David Thompson's Narrative 1784-1812*, edited with an introduction and notes by Richard Glover and published in 1962 by the Champlain Society.† Unlike Joseph Burr Tyrrell's 1916 edition of the *Narrative*, Glover's version contains a critical introduction to the man and his times.

---

† The Society publishes editions of historical journals and collections of documents relating to the history of Canada. With the exception of its Ontario Series (obtainable from the University of Toronto Press), the Society does not make its publications available to non-members. However, the Reprint Division of the Greenwood Press, Inc., 51 Riverside Avenue, Westport, Connecticut, 06880, has reprinted a facsimile edition of certain volumes published by the Champlain Society. Thus, for example, one can purchase, or request a library to purchase (or borrow), W. Stewart Wallace's *Documents Relating to the North West Company* or J. B. Tyrrell's 1916 edition of Thompson's *Narrative*.

GEORGE SIMPSON

There is no definitive biography of the Overseas Governor of the Hudson's Bay Company. However, a very readable portrait of Simpson as HBC leader, womanizer, statesman, and capitalist is sketched in John S. Galbraith's *The Little Emperor*, published by Macmillan of Canada in 1976.

ROBERT CAMPBELL

For reasons not very clear to me, there are many unrecognized heroes and heroines in Canadian history. Campbell was for long one of them but was lucky enough to be rescued from oblivion by Clifford Wilson in *Campbell of the Yukon,* published in 1970 by Macmillan of Canada.

THE FUR TRADE

For my money, the most readable, short account of fur trading in the Mackenzie watershed is contained in a little-known book entitled ***TRADER KING****, as told to Mary Weekes: the thrilling story of forty years' service in the North-West Territories, related by one of the last of the old time* Wintering Partners *of the Hudson's Bay Company*, Regina and Toronto, School Aids and Text Book Publishing, 1947. These memoirs of Chief Trader William Cornwallis King are leavened by great humor and an admirable understanding of the virtues and vices of the human species.

Quite a bit of romantic nonsense has been written about the fur trade by enthusiastic amateurs. On the other hand, professional historians have produced somewhat tedious tomes of use only to other historians. I have therefore compiled a list of those titles that I think are either important or popular enough to be of help and interest to the general reader.

The Course of Empire, by Bernard De Voto
Fur Trade Canoe Routes of Canada/Then and Now, by Eric W. Morse
The Gentlemen Adventurers, by R. E. Pinkerton
The Honourable Company: a history of the Hudson's Bay Company (rev. ed., 1949) by D. Mackay
**The Indian Heritage of America*, by A. M. Josephy, Jr.
Montreal and the Fur Trade, by E. E. Rich

The North West Company, by M. W. Campbell
The Pedlars from Quebec, and Other Papers on the Nor'Westers, by W. S. Wallace.
The Voyageur, by G. L. Nute.

THE MACKENZIE WATERSHED

There's a very extensive literature on this subject but, again, I've presumed to pick out only those books that I think will give the general reader a mixture of information and pleasure.

As Long As This Land Shall Last: A History of Treaty 8 and Treaty 11, 1870-1939, by René Fumoleau, OMI
The Big Dam Country: A Pictorial Record of the Development of the Peace River Country, by Bruce Ramsey and Dan Murray
Canada's North, by R. A. J. Phillips
Finlay's River, by R. M. Patterson
Moratorium: Justice, Energy, the North, and the Native People, by Hugh and Karmel McCullum and John Olthuis
Northern Frontier, Northern Homeland: The Report of the Mackenzie Valley Pipeline Inquiry: Volume One, by Commissioner Mr. Justice Thomas R. Berger
NORTHERN REALITIES: Canada-U.S. Exploitation of the Canadian North, by Jim Lotz
NORTH of 55°: Canada from the 55th Parallel to the Pole, edited by Clifford Wilson
Paddle Wheels to Bucket-Wheels on the Athabasca, by J. G. MacGregor
The Past and Future Land: An account of the Berger Inquiry into the Mackenzie Valley pipeline, by Martin O'Malley
Peace River Chronicles, selected and edited by Gordon E. Bowes
Ribbon of Water and Steamboats North: Meeting Place of Many Waters, Part Two in a History of Fort McMurray, 1870-1898, by D. J. Comfort
Seven Rivers of Canada [The Mackenzie, the St. Lawrence, the Ottawa, the Red, the Saskatchewan, the Fraser, the St. John], by Hugh MacLennan
Sixty Below, by Tony Onraet

**This Land Is Not For Sale*, by Hugh and Karmel McCullum
**Trail to the Interior*, by R. M. Patterson
The Yukon and Northwest Territories, [Macmillan's "The Traveller's Canada" Series] by Edward McCourt

OIL AND GAS

Books that explain, clearly and concisely, the basics of petroleum engineering, or production, or marketing, or politics are hard to come by. But a few of these are as follows:

**The Big Tough Expensive Job: Imperial Oil and the Canadian Economy*, edited by James Laxer and Anne Martin
The Brotherhood of Oil: Energy Policy and the Public Interest, by Robert Engler
**Fuelling Canada's Future*, by Wade Rowland
**Sellout: The Giveaway of Canada's Energy Resources*, by Philip Sykes
**THE SEVEN SISTERS: The Great Oil Companies and the World They Shaped*, by Anthony Sampson
**The Tar Sands: Syncrude and the Politics of Oil*, by Larry Pratt

Index

Collective entries have been made under the following headings: Fur depots and posts; Indian groups; Lakes; Rivers.

Abasand Oils Limited, 151
Aberhart, William, 150
Alaska, 180-183, 217
Alaska Highway, 144
Alberta, 138, 150, 152-153, 171-172, 177
Alberta Energy Company, 160
Alberta Gas Trunkline Company Ltd., 177
Alberta Research Council, 152
Alberta tar sands, 147-148, 153
Aleutian Islands, 144
Aluminex Limited, 166, 167
Alyeska Pipeline Service Company, 181-183
Anderson, James, 119
Andrew, Paul, 223-224
Arctic, 93
Arctic Circle, 93
asphalt, 132, 134
Astor, John Jacob, 61
Athabasca country, 61, 62, 65, 94
Athabasca Falls, 6
Athabasca Glacier, 5
Athabasca Landing, 104
Athabasca Pass, 46, 51
Athabasca tar sands, 3, 131, 148-152, 154-161, 165, 171-173
Atkinson Point, N.W.T., 176
Atlantic Richfield Canada, 159, 167, 175, 180
Atlantic Richfield Company, 175, 180
aviation, 141-143

"babiche people," 21, 95
Ball, Max W., 151
barrens, 94
Beaufort Sea, 3, 92, 134, 176, 242
beaver, 95-96, 99, 198
beaver hat, 55
Bechtel Corporation, 152, 182
Berger Report, *see Northern Frontier, Northern Homeland*
Berger, Thomas R., 177, 217-222, 239-245
Bering Strait, 193
bitumen, 132, 147-148
Bitumount, Alta., 148, 152
Black, Samuel, 38-42, 81, 90
Blair, Robert, 234, 235
Blair, Sydney M., 152, 153
Blake, Phillip, 224-226
Blondeau, Maurice, 58
Blondin, Gina, 201-202
bourdigneaux, 52
bourgeois, 70
British Columbia, 36, 37, 42, 112, 177
British Petroleum, 162, 180, 187, 189
Brown, John George "Kootenai," 137

Buchanan, Judd, 220
Butler. William, 42

Cameron, Agnes Deans, 109-110
Campbell, Robert, 113-121
Canada Cities-Service, 159, 167-168
Canada Development Corporation, 178
Canadian Arctic Gas Pipeline Limited, 177-178, 190, 214, 215, 218-219, 220, 235
Canadian Bechtel, 152
Canadian Petroleum Association, 174, 184, 186
Candle Oil, 166, 167
Canoe Encampment, B.C., 46
Canol project, 144-145
canot de maître, 69
canot du nord, 69
caribou, 198, 241
Cassiar Mountains, B.C., 40
castoreum, 96
Champion, L. R., 150
Charlie, (Mrs.) John, 226
Charlie, Martha John, 223
Cheadle, Walter Butler, 49-50
Chevron Standard, 167
Chief Commissioner, 128
Chief Factor, 88, 123, 130
"Chieftainess of the Nahanies," 115-116, 117
Chief Trader, 88, 123, 130
Chrétien, Jean, 213, 218
Church organizations, 198-201, 202, 214
Clark, Karl A., 152
Clarke, John, 75, 76, 78-79, 91
clerk, 36, 68
Cold Lake oil sands, 148
Columbia Icefield, 5
Colvile, Andrew, 82
Committee for Original Peoples Entitlement, 219
Conroy, Henry A., 210-211
continental divide, 45-46
Cook, James, 17, 63
"Cook's River," 17-18, 120, 121
"country wife," 97
Courneyer, Joseph, 42
Creole Petroleum, 163, 164

Dawson, Y. T., 139
Decoux, Father, 142
deed polls, 123, 127, 128, 130
Dene, 195-197, 200-203, 208-209, 210-212, 213-214, 216-217, 235-236
Dene Declaration, 235-236
Department of Energy, Mines and Resources, 171, 184, 220
Department of Indian Affairs and Northern Development, 177, 202-204, 213, 216-217, 218, 220
Derbyshire, Pete, 142
Douglas, David, 47-48
Douglas, James, 112
Drake, "Colonel" Edwin, 136
Drummond, Thomas, 47
Ducette, Charles, 30

Ellice, Edward "Bear," 122, 123-124
Ells, Sydney C., 145, 146, 147, 151
energy crisis, 174-175
engagé, 69
English Chief, 17, 22
Erasmus, Georges, 195, 235-236
Eskimo, *see* Inuit
Esso Research and Engineering, 159
Exxon Corporation, 159, 162, 175, 189

factor, 47
Finlay Forks, B.C., 32
Finlay & Gregory, 36, 57
Finlay, James, 36
Finlay, "Jocko," 44
Finlay, John, 36-38, 44
Fitzsimmons, R. C., 148-151
Foothills Pipe Lines Limited, 177, 214-215, 235
Forget, Amédeé Emmanuel, 208
Fort Fitzgerald, N.W.T., 139
Fort Franklin, N.W.T., 222
Fort McMurray, Alta., 105, 145-147, 215
Fort Smith, N.W.T., 108, 210, 222
Fraser, Simon, 37, 42, 43, 46
French, Layfayette, 137
Fullerton, Elmer, 143
Fumoleau, René, 211
Fur depots and posts: Churchill Factory, 26; Fort Chipewyan, 16, 28, 79, 107; Fort Edmonton, 48, 52; Fort Fork, 29; Fort Franklin, 222; Fort Garry, 119; Fort Good Hope, 109, 235; Fort Norman, 91; Fort Providence, 109; Fort Reliance, 109; Fort Resolution, 109; Fort St. John, 37, 207; Fort Selkirk, 118, 142; Fort Simpson, 109, 110, 118; Fort Vancouver, 48; Fort Wedderburn, 78, 79, 82; Fort William, 80, 89; Fort Yukon, 118; Grand Portage, 26, 59, 89; Jasper House, 48, 51; Michilimackinac, 59; Rocky Mountain House (B.C.), 37; York Factory, 26
Furlong, Charlie, 222-223

Genelli, Rainer, 229
Geological Survey of Canada, 206-207
Gillespie, Alastair, 185, 186
Gorman, George, 142
Goyer, Jean-Pierre, 166
Grahame, S.S., 107-108
Grant, Cuthbert, 38
Great Canadian Oil Sands, 154-157, 190
Green, Jim, 236-238
Greene, Joseph, 184, 186
Gregory, John, 36
Gregory & McLeod, 36, 67
Gulf Oil Canada, 159, 167, 176, 178, 187, 189
Gulf Oil Corporation, 159, 162

Hay River, N.W.T., 215
Hearne, Samuel, 9-10, 22, 94
Henday, Anthony, 56-57
Henry, Alexander (the Elder), 58, 59, 61-62
Henry, Alexander (the Younger), 101
Hill, Bill, 142
Hodgson, Stuart, 235
Home Oil, 166, 167
hommes du nord, 69
Horte, Vernon, 190
Howse Pass, 44
Hudson's Bay Company, 4, 26-27, 35, 46, 54, 56-57, 61, 71-72, 74, 75-76, 80-81, 82, 83-84, 87, 88-91, 96-109, 110-130, 202
Hudson's Bay Oil and Gas, 166, 167
Humble Oil and Refining, 175, 180

Imperial Oil Limited, 139, 141, 143, 144, 159, 164, 167, 176, 178, 186-187, 188, 209

Indian Brotherhood of the Northwest Territories, 219, 235
Indian groups: Beaver, 23, 28, 29, 33, 95, 197, 209; Blackfoot, 44; Carrier, 23; Chilkat, 118-119; Chipewyan, 21-22, 61-62, 95, 197, 209; Copperknives, 15-16, 22, 95; Cree, 28, 43, 95, 197, 209; Dogrib, 19-20, 22, 27-28, 95, 197; Hare, 22, 45, 197; Iroquois, 43, 48; Kootenay, 44; Loucheux (Kutchin), 22; Nipissing, 43; Nishga, 217; Piegan, 44; Sekani, 23, 34, 197; Slave, 19-20, 22, 28, 95, 197
"Inland Empire," 46
Inspecting Chief Factor, 128
International Bitumen Company, 149-150, 154
International Finance Society, 125, 126
Inuit, 23-24, 94, 194, 195, 202-203, 213-214, 216-217, 227-229

Johnson, William, 111
Johnston, Walter, 142
Junior Chief Factor, 128

Kahn, Herman, 164-165, 166
Kane, Paul, 26, 50-54

Lac la Martre, N.W.T., 236
La Guarde, Joseph, 42
Laird, David, 208
Lakes: Arctic, 34; Athabasca, 1, 7, 92; Great Bear, 3, 23, 92, 143; Great Slave, 1, 10, 15, 92; Thutadé, 8, 41
Landry, Joseph, 30
La Prise, Baptiste, 42
Leduc, Alta., 137, 138
Leroux, Laurent, 16
Lewis, David, 188-189
Link, Theodore August, 139-141, 143
Lougheed, Peter, 159-160, 161
Lysyk, Kenneth, 242

Macdonald, Donald, 165, 184-185
MacFarlane, Roderick Ross, 108, 129
Mackenzie, Alexander: authorship, 13-14, 20, 34-35; character, 10-12; early years, 36, 46, 57; explorations (1789), 13-25, (1793) 30-34; later years, 72-75; place in history, 34-36
Mackenzie Delta, 24-25, 132, 175-176, 178, 179, 242
Mackenzie Mountains, N.W.T., 18, 144
Mackenzie River, S.S., 109
Mackenzie Valley pipeline, 166, 183-184, 214-217, 242-244
Mackenzie Valley Pipeline Inquiry, 177, 219-245
"Made Beaver," 99
mal de racquet, 53
mangeurs du lard, 69
Manning, Ernest, 152-153, 154
Manson, Donald, 42
Mattonabee, 22
McConnell, R. G., 206-207
McCullum, Hugh and Karmel, 214
McDougall Pass, 194
McGillivray, Simon, 122, 123
McGillivray, Simon, Jr., 84
McGillivray, William, 72, 84, 89, 122, 123

McIvor, D. K., 186
McKay, Alexander, 30
McKenzie Commission, 162
McKenzie, Roderick, 16
McLeod, Alexander Roderick, 78
McLeod, Archibald Norman, 79, 81
McLoughlin, John, 112
McTavish, Frobisher & Co., 72
Melville Island, N.W.T., 148
Meredith, Edmund A., 205
Methye Portage, Alta., 62
Métis, 42, 48, 50, 98, 200, 203, 205, 213-214, 216-217
Métis Association of the Northwest Territories, 219, 235
Milton, Viscount William Fitzwilliam, 49-50
Moberly, John, 145
Mobil Oil Corporation, 162, 189
Montreal, 61
Morrow, William G., 211-212
Mount Edith Cavell, Alta., 45
Murphy Oil, 166, 167
muskrat, 96

National Energy Board, 167
National Oil Policy, 163
natural gas, 135, 137, 138, 176, 183-184, 188-192
New Caledonia (B.C.), 42
Norman Wells, N.W.T., 139-141, 143-144, 209
Northern Frontier, Northern Homeland, 239-244
Northern Transportation Company, 215
North West Company, 12, 26, 35, 36, 46, 58, 65-68, 71-72, 80-81, 87, 121-124, 130
Northwest Territories, 179, 202, 203, 204, 207, 210
North-West Navigation Company Limited, 107
Nor'Westers, 58

O'Byrne, Eugene Francis, 49-50
Ogden, Peter Skene, 38, 39, 91
oil and gas, 133-135, 137-139, 162-164, 173-175, 184-192
Oil Sands Limited, 150, 152
oil shales, 153
Oil Springs, Ont., 135, 136
Ojibwa, *see* Indian groups: Nipissing
Old Crow, Y.T., 222
Ontario, 169, 170, 177
Organization of Petroleum Exporting Countries, 175, 185

Pacific Lighting Gas Development Company, 178, 188
Pacific Western Airlines, 216
PanCanadian Petroleum, 189
Patrick, Alf, 140
Patrick, Russell, 154-155
Paulatuk, N.W.T., 222
pays d'en haut, 58
Peace Point, Alta., 28
Peace River, Alta., 29, 143
Peace River Canyon, 8, 30-32
Peace River country, 76-77
Peace River oil sands, 148
"Pedlars from Montreal," 58, 61
pemmican, 30, 77
permafrost, 93, 179-180
Perreault, Antoine, 42
Petro-Canada, 157
Petrofina Canada, 166, 167
petroleum, *see* oil and gas; *see also* natural gas

Pew, John Howard, 154-155
pingos, 179
Polar Gas Limited, 188, 189
Pond, Peter, 60-63
Portage Mountain, B.C., 30, 133
Port Radium, N.W.T., 143
Prudhoe Bay, Alaska, 175, 180-181, 183
Public Petroleum Association of Canada, 187

Ramparts, 23, 132
Red River Settlement, 74, 114
reservoir rocks, 134
Rivers: Assiniboine, 56; Athabasca, 1, 3, 6-7, 26, 42, 45, 46, 48-53, 63, 92, 154, 157; Bella Coola, 34; Churchill, 62; Clearwater, 62; Columbia, 46; "Cook's River," 17-18, 120, 121; "Disappointment," 25; Finlay, 32, 36, 37, 40-41; Great Bear, 23; Jean, 13; Liard, 1, 18, 92, 114-115, 117, 118; Mackenzie, 1-3, 6, 17-25, 41, 63, 91, 92, 132-133, 179; Miette, 45; Mississippi-Missouri, 2, 25; Pack, 34; Parsnip, 32, 33-34; Peace, 1, 8-9, 10, 26, 28, 30, 32, 63, 76-77, 92; Peel, 118, 121; Pelly, 118; Porcupine, 118, 121, 194; Rochers, 7; Saskatchewan, 46; Slave, 1, 9-10; Smoky, 29; Sunwapta, 6, 10; Whirlpool, 45; Yukon, 28, 118, 121
Robertson, Colin, 76, 98, 122
Rocky Mountains, 8
Rocky Mountain Trench, 40
Routledge, W. H., 146
Rowand, John, 112
Royal Canadian Mounted Police, 146, 202, 207
"rubaboo," 69
Rupert's Land, 77, 126

St. Laurent, Louis, 202
sault, 69
scowmen, 105-107
sea otter, 28
Selkirk, Thomas Douglas, 5th Earl of, 74-76, 80, 122
"Seven Sisters," 162-163
Sharp, Mitchell, 219
Shell Canada, 166, 178
Shell Explorer, 166-167
Shell Oil, 162, 187, 189
Simcoe, John Graves, 72
Simpson, George, 35, 39, 81-91, 96-98, 102, 114, 118, 120
Sir Alexander Mackenzie & Co., *see* XY Company,
Smith, Donald, 108
Smith, John H., 107-108
Smith, William A., 136
Smith's Landing, N.W.T., 139
Southesk, James Carnegie, 9th Earl of, 48-49
Standard Oil Company of California, 162, 189
Standard Oil Company of New Jersey, *see* Exxon Corporation
Steen, Vince, 227-229
Stikine Mountains, B.C., 40
Stirling, Elleonora, C., 120, 121
Sun Oil Company, 154-155
Syncrude Canada, 159-161, 164, 166, 167-170

tailings, 156
Tanner, Nathan E., 150
Tarrangeau, Jean Baptiste, 42
tar sands, 146; *see also* Alberta tar sands

Taylor, Charles, 141
Texaco (Canada), 187
Texaco Incorporated, 162, 189
Texas Eastern Transmission Corporation, 178, 188
"Thomas the Iroquois," 43, 45
Thompson, David, 43-46, 59
Titusville, N.Y., 136
TransCanada PipeLines Limited, 178, 188
treaties, 205-212, 213-214
"tree line," 93-94
tripe de roche, 59
Trudeau, Pierre Elliott, 165, 169, 170, 176, 188
T'Seleie, Frank, 233-235
Turner, John, 168, 169
Turner Valley, Alta., 137, 138, 141

Usher, Peter, 229-233

Valdez, Alaska, 176
Vancouver Island, 112
Venezuela, 148
Voyages from Montreal, 11, 35, 73
voyageur, 68-71

Wabasca oil sands, 148
Wallace, J. N., 38
Walters, John, 107
Wedderburn, Andrew, *see* Colvile, Andrew
West Coast Transmission, 177
"wet gas," 137
Whitehorse, Y.T., 144, 229
Wilder, William, 190
Williams, James M., 135, 136
Williams, William, 82, 83
Willson, Bruce, 168-169, 191-192
wintering partner, 65, 68
wolverine, 98-99
Wrigley, Joseph, 129
Wrigley, S.S., 108-109
Wyllie, 110-111

XY Company, 73

Yellowknife, 144, 219
Yellowknife Bay, 15
York boat, 101-103
Youel, Christiana, 36
Yukon Territory, 179, 202, 208, 241-242